ELECTRON SPIN RESONANCE
OF METAL COMPLEXES

ELECTRON SPIN RESONANCE
OF METAL COMPLEXES

Proceedings of the Symposium on ESR of Metal Chelates at the
Pittsburgh Conference on Analytical Chemistry and Applied
Spectroscopy, held in Cleveland, Ohio, March 4-8, 1968

Edited by
TEH FU YEN
Department of Chemistry
California State College at Los Angeles

ℚ PLENUM PRESS • NEW YORK • 1969

Library of Congress Catalog Card Number 69-19169

© 1969 Plenum Press, New York
A Division of Plenum Publishing Corporation
227 West 17th Street, New York, N. Y. 10011

Foreword

For a number of years, there existed a real gap between the science of metal complexes and that of electron spin resonance (ESR). Simple reasons account for this fact. At a certain stage of development the scientists engaged in investigations of metal complexes did not have access to ESR instrumentation, while on the other hand, ESR theoreticians rarely had an interest in exploring the chemical applications of metal complexes. More recently chemical physicists have started to make intensive efforts to bridge the gap by applying the ESR technique to a wide range of chemical problems, particularly those involving transition metals and their complexes. In large measure the successes of the theory of the electronic structure of transition metal ions are due to the comprehensive and precise results of ESR studies by chemical physicists. On the other hand, chemists also seem to have realized lately that an immense amount of information can be obtained from ESR data.

It is obvious, therefore, that a symposium bringing together the various disciplines was necessary, and there was little doubt that in such a symposium a considerable advantage could be gained from the exchange of information among scientists with different research interests. Consequently, a symposium on "ESR of Metal Chelates" was held on March 4, 1968, at the Pittsburgh Conference on Analytical Chemistry and Applied Spectroscopy, at the Cleveland Convention Center. The participants in the symposium did indeed reach the conclusion that the objectives of cross-breeding and fertilization among different disciplines had been achieved to a significant extent. The success of the meeting definitely warranted the publication of the lectures that were presented. In order to obtain a truly representative coverage of the field, three additional invited papers have been included in this volume.

The contents of this monograph give a wide scope of the main problems in ESR of metal complexes, e.g., the structure and bonding of the transition metals and their ions; the charge transfer, symmetry reduction, and ligand type of their chelates; computer synthesis of spectra; and high-resolution techniques. The topics discussed also involve organometallic chemistry, solid state physics, biomedical science, instrumentation, geoscience, and computer programming. This broad coverage and diversity of subjects contribute to the value of this volume.

Acknowledgments are due in the first place to Dr. J. O. Frohliger, Program Chairman of the 1968 Pittsburgh Conference, for his encouragement, to

MAR 25 1970

many of the editor's colleagues, coworkers, and friends for their suggestions, and above all to the active participants for their valuable contributions.

Teh Fu Yen

Los Angeles, California
May, 1969

Contents

Contributors

R. Linn Belford, Noyes Chemical Laboratory, University of Illinois, Urbana, Illinois

W. E. Blumberg, Bell Telephone Laboratories, Murray Hill, New Jersey

L. J. Boucher, Department of Chemistry, Carnegie–Mellon University, Pittsburgh, Pennsylvania

Sunney I. Chan, Arthur Amos Noyes Laboratory of Chemical Physics, California Institute of Technology, Pasadena, California

H. M. Gladney, IBM Research Laboratory, San Jose, California

Robert G. Hayes, Department of Chemistry, University of Notre Dame, Notre Dame, Indiana

Michael A. Hitchman, University of Illinois, Urbana, Illinois

B. Johnson, IBM Research Laboratory, San Jose, California

Nobuko Kataoka, Laboratory of Physical Biology, National Institute of Arthritis and Metabolic Diseases, Bethesda, Maryland

Hideo Kon, Laboratory of Physical Biology, National Institute of Arthritis and Metabolic Diseases, Bethesda, Maryland

Marcel Kopp, Mellon Institute, Carnegie–Mellon University, Pittsburgh, Pennsylvania

John H. Mackey, Mellon Institute, Carnegie–Mellon University, Pittsburgh, Pennsylvania

P. T. Manoharan, Department of Chemistry, Michigan State University, East Lansing, Michigan

B. R. McGarvey, Department of Chemistry, Polytechnic Institute of Brooklyn, Brooklyn, New York

Brian W. Moores, Noyes Chemical Laboratory, University of Illinois, Urbana, Illinois

J. Peisach, Department of Pharmacology and Molecular Biology, Albert Einstein College of Medicine, Yeshiva University, Bronx, New York

Max T. Rogers, Department of Chemistry, Michigan State University, East Lansing, Michigan

Louis D. Rollman, Arthur Amos Noyes Laboratory of Chemical Physics, California Institute of Technology, Pasadena, California

J. D. Swalen, IBM Research Laboratory, San Jose, California

Edmond C. Tynan, Radiation Research Laboratories, University of Notre Dame, Notre Dame, Indiana

David E. Wood, Department of Chemistry, Carnegie–Mellon University, Pittsburgh, Pennsylvania

Teh Fu Yen, Department of Chemistry, California State College, Los Angeles, Los Angeles, California

Charge Transfer in the Metal–Ligand Bond as Determined by Electron Spin Resonance

B. R. McGarvey

Department of Chemistry
Polytechnic Institute of Brooklyn
Brooklyn, New York

A self-consistent charge analysis has been performed on the ESR spin-Hamiltonian parameters of CrO_4^{3-} and MoO_4^{3-} in which values of the spin-orbit constant and $\langle r^{-3} \rangle$ for different charges on the ion are used until the charge calculated from the molecular orbitals is consistent with that assumed. In both cases the effective charge on the metal ion is found to be $+2$ or less. The effects of several approximations on the calculation are tested, and an approximate analysis is made of the charge transfer in other M^{5+} and M^{4+} ions with the d^1 configuration.

INTRODUCTION

In deriving MO coefficients from the ESR spin Hamiltonians of transition-metal complexes the orbital spin-orbit parameter ξ, the value of $\langle r^{-3} \rangle$, and the energies of several excited d states must be known. Appreciable covalency in the metal–ligand bond can make the effective charge on the metal atom much different from the formal oxidation number, and this makes the choice of the proper values for ξ and $\langle r^{-3} \rangle$ difficult due to their strong dependence on charge.

Little difficulty was encountered in studies on Cu(II) complexes[1-5] by using the Cu^{2+} values of ξ and $\langle r^{-3} \rangle$, because ξ varies very little for Cu°, Cu^+, and Cu^{2+}. Also, as will be discussed further on, the effective charge in the nearly-filled d level can change very little with covalency. In the case of VO(IV) complexes, however, Kivelson and Lee[6] found it necessary to use V^{2+} values of ξ and $\langle r^{-3} \rangle$ in order to arrive at any sensible results. In other studies[7-11] these parameters have been chosen by various criteria.

A knowledge of the relevant MO coefficients for the d orbitals allows the calculation of the effective charge on the metal atom, and a proper approach to the problem would be a self-consistent calculation in which the values of ξ and $\langle r^{-3} \rangle$ are varied until the charge computed from the MO coefficients matches that assumed in choosing ξ and $\langle r^{-3} \rangle$. The ion CrO_4^{3-}, which has been studied in this laboratory,[12] appears to be an ideal system for the application of this approach, as information about all the d orbitals can be obtained from the spin Hamiltonian and good values of ξ and $\langle r^{-3} \rangle$ are available for most charges of the free ion. Furthermore, the high oxidation number of $+5$ for this ion would lead us to expect considerable charge transfer in the Cr—O bonds.

PRELIMINARY ANALYSIS

Before attacking the MO problem a preliminary analysis of the data must be considered to justify some of the assumptions which will be made. The spin-Hamiltonian parameters for CrO_4^{3-} in Ca_2PO_4Cl are

$$g_z = 1.9936, \qquad g_x = g_y = 1.9498$$

$$A_z = (+)7.6 \times 10^{-4} \, cm^{-1}$$

$$A_x = (-)22.7 \times 10^{-4} \, cm^{-1}$$

$$A_y = (-)19.1 \times 10^{-4} \, cm^{-1}$$

$$\tfrac{1}{2}(A_x + A_y) = (-)20.9 \times 10^{-4} \, cm^{-1}$$

These parameters are basically the type expected for an electron in a d_{z^2} orbital and indicate that the tetrahedral symmetry is distorted to bring the d_{z^2} orbital lowest in energy. The d_{xy} orbital, which is degenerate with the d_{z^2} orbital in tetrahedral symmetry, is energetically not far from the d_{z^2} orbital, as evidenced by the fact that T_1 is short enough at room temperature to make the ESR spectrum undetectable. Thus the distortion from tetrahedral symmetry is small.

For a d_{z^2} orbital g_z should be 2.0023 to second order in perturbation theory. The difference between g_z and 2.0023 could be explained by going to higher-order perturbation theory, provided that $\xi/\Delta E_{xz,yz}$ is large. However, this parameter is at most 0.03, and cannot account for the major part of the difference.

The differences between g_z and 2.0023 and between A_x and A_y are due to a small distortion from axial symmetry. In C_{2v} symmetry both the d_{z^2} and $d_{x^2-y^2}$ orbitals belong to A_1 and the z^2 and $x^2 - y^2$ orbitals must be written as

$$\psi_{z^2} = ad_{z^2} + bd_{x^2-y^2}$$
$$\psi_{x^2-y^2} = ad_{x^2-y^2} - bd_{z^2} \tag{1}$$

Assuming pure d orbitals and using first- and second-order perturbation theory, we obtain the following equations for the spin-Hamiltonian parameters:

$$g_z = 2.0023 - \frac{8b^2\xi}{\Delta E_{xy}}, \qquad g_x = 2.0023 - \frac{2\xi(\sqrt{3}a + b)^2}{\Delta E_{yz}},$$

$$g_y = 2.0023 - \frac{2\xi(\sqrt{3}a - b)^2}{\Delta E_{xz}}$$

$$A_z = -K + P\left\{ \frac{4}{7}(a^2 - b^2) + \frac{\xi(\sqrt{3}a - b)(\sqrt{3}a + 3b)}{7\,\Delta E_{xz}} \right.$$

$$\left. + \frac{\xi(\sqrt{3}a + b)(\sqrt{3}a - 3b)}{7\,\Delta E_{yz}} - \frac{8b^2\xi}{\Delta E_{xy}} \right\}$$

(2)

$$A_x = -K + P\left\{ -\frac{2}{7}(a^2 - b^2) - \frac{4\sqrt{3}}{7}ab + \frac{4\sqrt{3}\xi ab}{7\,\Delta E_{xy}} \right.$$

$$\left. - \frac{\xi(\sqrt{3}a - b)(\sqrt{3}a + 3b)}{7\,\Delta E_{xz}} - \frac{2\xi(\sqrt{3}a + b)^2}{\Delta E_{yz}} \right\}$$

$$A_y = -K + P\left\{ -\frac{2}{7}(a^2 - b^2) + \frac{4\sqrt{3}}{7}ab - \frac{4\sqrt{3}\xi ab}{7\,\Delta E_{xy}} \right.$$

$$\left. - \frac{\xi(\sqrt{3}a + b)(\sqrt{3}a - 3b)}{7\,\Delta E_{yz}} - \frac{2\xi(\sqrt{3}a - b)^2}{\Delta E_{xz}} \right\}$$

$$P = 2.0023 g_N \beta_e \beta_N \langle r^{-3} \rangle$$

where K is the isotropic term. From the spin-Hamiltonian parameters of CrO_4^{3-} we find the values

$$b^2 = 0.0053, \qquad P = -31.6 \times 10^{-4}\,\text{cm}^{-1}, \qquad K = -10.2 \times 10^{-4}\,\text{cm}^{-1}$$

$$\xi/\Delta E_{xy} = 0.21, \qquad \xi/\Delta E_{xz} = 0.00959, \qquad \xi/\Delta E_{yz} = 0.00810$$

The small value of b^2 indicates that a very small admixture of $d_{x^2-y^2}$ accounts for the difference between A_x and A_y as well as the difference in g_z from 2.0023. The large value of $\xi/\Delta E_{xy}$ supports the T_1 observation that ΔE_{xy} is small and therefore the distortion from tetrahedral symmetry small. The value of P is close to that of the $+1$ ion given in Table I, and is a strong indication that there is considerable charge transfer in the Cr—O bond.

In C_{2v} the $3d_{z^2}$ and $4s$ orbital belong to the same irreducible representation A_1, and hence can be mixed. We can deduce the extent of this admixture from the value of K. The value of χ as defined by Abragam et al.,[13]

$$\chi = (4\pi/S)\left(\psi \left| \sum_i \delta(r_i)s_{zi} \right| \psi \right)$$

(3)

is calculated to be -1.53 from K. Normal values of χ for Cr(V) complexes, in which symmetry does not allow $4s$ admixtures in the ground state is -2.6.[14] Using Watson's[15] wave function for $4s$, the value of χ for a $4s$ electron would be $+389$. Thus the $4s$ contribution to χ is $+1.1$, and the per cent $4s$ character is 0.3%.

TABLE I
Values of ζ and P for Chromium Ions

Ion	ζ, cm$^{-1(a)}$	$10^4 P$, cm$^{-1(b)}$
Cr^{5+}	383	(50.2)
Cr^{4+}	342	45.0
Cr^{3+}	276	39.7
Cr^{2+}	234	34.6
Cr^{+}	(232)	(29.5)
Cr$^{\circ}$	230	24.4

[a]Computed from the spectroscopic tables of C. E. Moore, Atomic Energy Levels, National Bureau of Standards Circular 467, Vol. 1 (1949); Vol. 2 (1952); Vol. 3 (1958).
[b]Computed from the values of $\langle r^{-3} \rangle$ obtained by A. J. Freeman and R. E. Watson, in: *Magnetism*, Vol. IIA, G. T. Rado and H. Suhl, eds. (Academic Press, New York, 1965), p. 167.

Although the preceding analysis is based on equations which assume a crystal-field model, it is clear from the results that in an MO analysis it will be a good approximation to treat the system as a simple tetrahedral case and ignore the small distortions that lead to small admixtures of $d_{x^2-y^2}$ and $4s$ into the ground state. This means that we can take the MO coefficients of the d_{xy} orbital to be the same as the d_{z^2} orbital and the coefficients of the $d_{x^2-y^2}$ orbital to be the same as those for the d_{xz} and d_{yz} orbitals. Thus we will need to find only two MO coefficients from the three pertinent spin-Hamiltonian parameters of $g_\perp = \frac{1}{2}(g_x + g_y)$, $A = A_z$, and $B = \frac{1}{2}(A_x + A_y)$.

MO ANALYSIS

The coordinate axes used are shown in Fig. 1. The z axis for each ligand atom is taken to be parallel to the bond direction between the metal atom and the ligand atom. The molecular orbitals are written

$$A_1 = \beta d_{z^2} - \tfrac{1}{2}\beta'(p_x^1 - p_x^2 - p_x^3 + p_x^4)$$

$$E_{xz} = \alpha d_{xz} + (\alpha'/\sqrt{2})(\cos \omega)(p_z^1 - p_z^4)$$

$$- (\alpha'/2\sqrt{2})(\sin \omega)(p_x^1 - p_x^4 - \sqrt{3}p_y^2 + \sqrt{3}p_y^3) \qquad (4)$$

$$E_{yz} = \alpha d_{yz} - (\alpha'/\sqrt{2})(\cos \omega)(p_z^2 - p_z^3)$$

$$-(\alpha'/2\sqrt{2})(\sin \omega)(p_x^2 - p_x^3 - \sqrt{3}p_y^1 + \sqrt{3}p_y^4)$$

In constructing these orbitals, the ligand s orbitals have not been included, as they would require an additional coefficient without contributing much

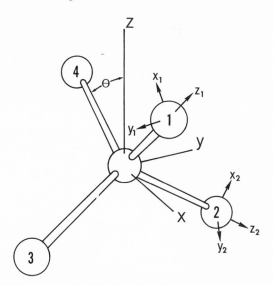

Fig. 1. Coordinate axes used in MO calculations on
CrO_4^{3-}.

to the problem. Also, there is no reason to expect sp hybridization in the CrO_4^{3-} ion. The value of ω will be chosen to maximize the overlap integral. Using Eq. (4), the spin-Hamiltonian parameters are

$$g_\perp = 2.0023 - (6\xi/\Delta E_{xz})\{\alpha^2\beta^2 - \alpha\alpha'\beta^2 S_E - \alpha^2\beta\beta' S_A$$

$$-(6)^{-1/2}\alpha\alpha'\beta\beta'[(\cos\omega) - (\sqrt{3}/2)\sin\theta\sin\omega]\}$$

$$A = -K + P\{\tfrac{4}{7}\beta^2 + (6\xi/7\,\Delta E_{xz})\alpha^2\beta^2\}$$

$$B = -K + P\{-\tfrac{2}{7}\beta^2 - (45\xi/7\,\Delta E_{xz})\alpha^2\beta^2\} \qquad (5)$$

$$S_A = \tfrac{1}{2}\int d_{z^2}(p_x^1 - p_x^2 - p_x^3 + p_x^4)\,d\tau$$

$$S_E = [(\cos\omega)/\sqrt{2}]\int d_{yz}(p_z^2 - p_z^3)\,d\tau$$

$$+[(\sin\omega)/2\sqrt{2}]\int d_{yz}(p_x^2 - p_x^3 - \sqrt{3}p_y^1 + \sqrt{3}p_y^4)\,d\tau$$

In deriving these equations, the spin-orbit contribution of the ligand oxygen atoms has been ignored due to its smallness relative to that of the chromium atom. The connection with charge density becomes clearer if we use the Mulliken[16] definition

$$P_\sigma = \alpha^2 - \alpha\alpha' S_E, \qquad P_\pi = \beta^2 - \beta\beta' S_A \qquad (6)$$

where P_σ and P_π are the charge densities on the metal ion for an electron in the E and A_1 orbitals, respectively. Using Eq. (6) and ignoring the negligible S^2 terms, we can rewrite $g_\perp$ as

$$g_\perp = 2.0023 - \frac{6\xi}{\Delta E_{xz}}\left\{P_\sigma P_\pi - (6)^{-1/2}\alpha\alpha'\beta\beta'\left[(\cos\omega) - \frac{\sqrt{3}}{2}\sin\theta\sin\omega\right]\right\} \quad (7)$$

As has been pointed out previously for the Cr^{3+} case,[10,17] it may be necessary to consider the contribution of the excited state arising from the promotion of a bonding electron in the $^\ddagger E_{xz,yz}$ state to the A_1 ground state. The wave function for the $^\ddagger E_{yz}$ bonding orbital is

$$^\ddagger E_{yz} = {}^\ddagger\alpha\, d_{yz} + ({}^\ddagger\alpha'/\sqrt{2})(\cos\omega)(p_z^2 - p_z^3) + ({}^\ddagger\alpha'/2\sqrt{2})$$

$$(\sin\omega)(p_x^2 - p_x^3 - \sqrt{3}p_y^1 + \sqrt{3}p_y^4) \quad (8)$$

and the additional contribution to $g_\perp$ is

$$\Delta g_\perp = \frac{6\xi}{\Delta E_{xz}^\ddagger}\left\{P_\pi(1 - P_\sigma) + (6)^{-1/2\ddagger}\alpha^\ddagger\alpha'\beta\beta'\left[(\cos\omega) - \frac{\sqrt{3}}{2}\sin\omega\sin\theta\right]\right\} \quad (9)$$

and to A and B is

$$\Delta A = -P\{6\xi^\ddagger\alpha^2\beta^2/7\,\Delta E_{xz}^\ddagger\}, \qquad \Delta B = P\{45\xi^\ddagger\alpha^2\beta^2/7\,\Delta E_{xz}^\ddagger\} \quad (10)$$

In Eq. (9) use was made of the fact that orthogonality of $^\ddagger E_{yz}$ with E_{yz} requires the relationship

$$(1 - P_\sigma) = {}^\ddagger\alpha^2 + {}^\ddagger\alpha^\ddagger\alpha'S_E \quad (11)$$

The number of electrons in the metal d orbitals of the bonding σ orbitals is $6(1 - P_\sigma)$ and for the bonding π orbitals is $4(1 - P_\pi)$. Therefore the effective charge Q on the Cr^{5+} ion is

$$Q = -4 + 6P_\sigma + 3P_\pi \quad (12)$$

Equation (12) assumes no charge transfer into the $4s$ or $4p$ orbitals. Such charge transfer is unimportant to our case, since electrons in these orbitals do little to shield the d orbitals and hence have little effect on the values of ξ and $\langle r^{-3}\rangle$ of the d orbitals. Therefore Q in Eq. (12) represents the minimum of charge transfer possible in the system.

Values of overlap integrals were taken from Jaffe[18] and Jaffe and Doak[19] and the values of S_E and S_A used were $S_A = 0.104$ and $S_E = 0.177$, with $\omega = 20°$. For ΔE_{xz} there are two reasonable choices of 10,000 cm^{-1} and 17,000 cm^{-1} according to polarization studies of these crystals by Banks et al.[20] For $\Delta E_{xz}^\ddagger$ the 34,000-cm^{-1} band is the lowest possible value that can be considered as reasonable.

In Table II the results of calculations are presented using Eqs. (5)–(10) for various assumed charges and possible choices of ΔE_{xz} and $\Delta E_{xz}^\ddagger$. Also

TABLE II
MO Results for Cr^{5+} in Ca_2PO_4Cl

Assumed charge	ΔE_{xz}, cm^{-1}	$\Delta E_{xz}^{\ddagger}$, cm^{-1}	α^2	β^2	P_σ	P_π	Q
+2	10,000	∞	0.688	0.830	0.583	0.781	+1.841
+3	10,000	∞	0.722	0.715	0.617	0.660	+1.684
+4	10,000	∞	0.750	0.602	0.646	0.543	+1.505
+2	10,000	34,000	0.682	0.862	0.576	0.816	+1.904
+3	10,000	34,000	0.720	0.730	0.615	0.675	+1.715
+4	10,000	34,000	0.735	0.620	0.631	0.562	+1.472
+2	17,000	∞	0.931	0.866	0.849	0.820	+3.554
+3	17,000	∞	0.956	0.740	0.879	0.686	+3.332
+4	17,000	∞	0.952	0.635	0.874	0.578	+2.978
+2	17,000	34,000	0.975	0.864	0.904	0.818	+3.878
+3	17,000	34,000	0.990	0.738	0.924	0.684	+3.596
+4	17,000	34,000	0.987	0.634	0.919	0.577	3.245
For $S_A = S_E = 0$							
+2	10,000	∞	0.480	0.878	0.480	0.878	+1.514
+3	10,000	∞	0.509	0.746	0.509	0.746	+1.292
+4	10,000	∞	0.504	0.645	0.504	0.645	+0.959
+2	10,000	34,000	0.517	0.895	0.517	0.895	+1.787
+3	10,000	34,000	0.541	0.765	0.541	0.765	+1.541
+4	10,000	34,000	0.545	0.658	0.545	0.658	+1.244

included are results of calculations in which the overlap integrals were assumed to be zero. In Fig. 2 calculated values of Q are plotted against the assumed charge of the ion.

DISCUSSION OF RESULTS

It will be noted in Table II and Fig. 2 that Q changes slowly with the assumed charge of the ion, although the values of α^2 and β^2 change greatly. Furthermore, the effect of assuming $S = 0$ or $\Delta E_{xz}^{\ddagger} = \infty$ is small. For $\Delta E_{xz} = 10,000 \, cm^{-1}$ the assumed charge and Q become the same at a charge of $+1.9$, while for $\Delta E_{xz} = 17,000 \, cm^{-1}$ it occurs at a charge of $+3.3$–3.4. However, if we look at P_σ and P_π, we see that for $\Delta E_{xz} = 17,000 \, cm^{-1}$ the bonding at a charge of $+3$ would be predicted to be mostly π bonding, with little σ bonding. Since this is an unlikely state of affairs, it appears that $\Delta E_{xz} = 10,000 \, cm^{-1}$ is the better choice, as it predicts the expected situation of extensive σ bonding plus moderate π bonding.

Azarbayejani and Merlo[21] have reported the spin-Hamiltonian values for Mo^{5+} in $CaWO_4$, where the Mo^{5+} ion is in a tetrahedral site distorted so as to bring the d_{z^2} orbital lowest. The crystal fields in $CaWO_4$ and Ca_2PO_4Cl should be similar, since the g values of Cr^{5+} in $CaWO_4$[22] are

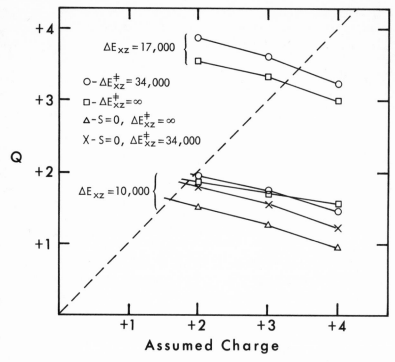

Fig. 2. Plot of the charge Q calculated from the molecular orbitals against the charge assumed in choosing ξ and $\langle r^{-3} \rangle$.

similar to those in Ca_2PO_4Cl. Since assumption of $S = 0$ and $\Delta E_{xz}^{\ddagger} = \infty$ has only a small effect on the value of Q, we can apply the MO equations to this system with these assumptions. In Table III are given the values of ξ

TABLE III
Values of ξ and P for Molybdenum Ions

Ion	$\xi \, cm^{-1}$	$10^4 P, \, cm^{-1}$
Mo^+	$672^{(a)}$	$-42.3^{(d)}$
Mo^{2+}	$745^{(c)}$	$-48.2^{(d)}$
Mo^{3+}	$817^{(a)}$	$-54.4^{(d)}$
Mo^{4+}	$887^{(a)}$	$-60.5^{(c)}$
Mo^{5+}	$1031^{(b)}$	$-66.7^{(c)}$

[a]J. S. Griffith, *The Theory of Transition Metal Ions* (Cambridge University Press, New York, 1961), p. 437.
[b]See footnote *a* of Table I.
[c]Interpolated or extrapolated.
[d]See footnote *b* of Table I.

and P for various molybdenum ions. In Table IV are given the results for assumed values of $\Delta E_{xz} = 10,000 \text{ cm}^{-1}$ and $20,000 \text{ cm}^{-1}$. Here again we must assume that ΔE_{xz} is closer to $10,000 \text{ cm}^{-1}$ than $20,000 \text{ cm}^{-1}$, since the larger value would predict large π bonding and little σ bonding. Therefore the charge on the Mo^{5+} ion is probably somewhere between $+1$ and $+2$, a result similar to that found for Cr^{5+}.

TABLE IV
MO Results for Mo^{5+} in $CaWO_4$

Assumed charge	ΔE_{xz}; cm^{-1}	P_σ	P_π	Q
$+1$	10,000	0.496	0.705	$+1.091$
$+2$	10,000	0.560	0.584	$+1.102$
$+3$	10,000	0.617	0.493	$+1.181$
$+4$	10,000	0.674	0.420	$+1.304$
$+1$	20,000	0.846	0.728	$+3.260$
$+2$	20,000	0.901	0.617	$+3.257$
$+3$	20,000	0.946	0.530	$+3.266$
$+4$	20,000	0.980	0.461	$+3.263$

An earlier analysis[23] of g values for the d^3 ions V^{2+}, Cr^{3+}, and Mn^{4+} in oxide lattices also showed that the effective charge in the d orbitals was $+2$ in all the ions. In this analysis, however, it was necessary to assume no π bonding because there was only one g value from which to calculate MO coefficients. Kivelson and Lee[6] have also found the effective charge of V^{4+} in VO(IV) complexes to be $+2$.

The charge analysis of ESR parameters used in this work cannot be applied to all systems. For example, in the VO^{2+} ion the MO coefficients of the d_{z^2} orbital cannot be determined from the spin Hamiltonian. This analysis is most important for ions with charges greater than $+2$ and d shells less than half-filled. For many ions in the first transition series ξ changes very little from 0 to $+2$. For ions with more than half-filled d shells the transfer of charge into d orbitals in the bonding orbitals is compensated by the transfer out of the d orbitals in the antibonding orbitals. In the case of Cu^{2+} the value of Q is given by

$$Q(Cu^{2+}) = 1 + P_\sigma(x^2 - y^2) \tag{13}$$

where $P_\sigma(x^2 - y^2)$ is the charge density on the copper ion for an electron in the antibonding $d_{x^2-y^2}$ orbital. Thus $Q(Cu^{2+})$ varies only between $+1$ and $+2$, and we see that little error is incurred by choosing the $+2$ values of ξ and P.

Although the analysis used here cannot be applied to many systems due to lack of information, we can deduce some information about the charge on the ion from the hyperfine terms. The value of β^2 is determined primarily from the values of A and B. Thus β^2 depends mainly on the choice of the value of P. If we know the correct value of P, we can obtain good values of β, the MO coefficient for the ground state, by using the equations

$$A = -K + P\{\tfrac{4}{7}\beta^2 + \tfrac{1}{7}(2.0023 - g)\}$$
$$B = -K + P\{-\tfrac{2}{7}\beta^2 + \tfrac{15}{14}(2.0023 - g)\} \tag{14}$$

when d_{z^2} is the ground state and

$$A = -K + P\{-\tfrac{4}{7}\beta^2 + (g - 2.0023) + \tfrac{3}{7}(g - 2.0023)\}$$
$$B = -K + P\{\tfrac{2}{7}\beta^2 + \tfrac{11}{14}(g - 2.0023)\} \tag{15}$$

when $d_{x^2-y^2}$ or d_{xy} is the ground state.

For d^1 configurations the ground state is always a π antibonding orbital, so that the overlap integral S is generally small and β^2 can be regarded as the electron's density on the metal ion. Since σ bonding should be greater than π bonding, a value of Q computed on the assumption that the charge density is the same for all orbitals will represent a maximum value of Q.

Values of β^2 for M^{5+} and M^{4+} ions have been calculated from Eqs. (14) and (15) using P's for various charges, and Q was then found from the equations

$$Q(M^{5+}) = -4 + 9\beta^2, \qquad Q(M^{4+}) = -5 + 9\beta^2 \tag{16}$$

This Q was plotted against the assumed charge and the charge noted at which they become the same. When this procedure is followed for Cr^{5+} in

TABLE V

Values of Q_{max} and β^2 for M^{5+} and M^{4+} Ions of Configuration d^1

Ion	Q_{max}	β^2	Ref.
Cr^{5+} in K_2CrOCl_5	+3.10	0.79	(a)
Mo^{5+} in K_2MoOF_5	+3.90	0.88	(b)
Mo^{5+} in $(NH_4)_2InCl_5 \cdot H_2O$	+3.45	0.83	(c)
Mo^{5+} in $K_3InCl_6 \cdot 2H_2O$	+3.25	0.81	(d)
Mo^{5+} in $K_2MoO(SCN)_5$	+2.40	0.71	(e)
Mo^{5+} in $K_3Mo(CN)_8$	+1.55	0.62	(f)
VO^{2+} in $Zn(NH_4)_2(SO_4)_2 \cdot 6H_2O$	+2.60	0.84	(g)
Nb^{4+} in $Nb(OCH_3)Cl_5^{2-}$	+2.10	0.79	(h)
Nb^{4+} in SnO_2	+2.40	0.82	(i)

^{a}H. Kon and N. E. Sharpless, *J. Chem. Phys.* **42**, 906 (1965).
^{b}N. S. Garifyanov, V. N. Fedotov, and N. S. Kucheryavenko, *Izv. Akad. Nauk SSSR, Ser. Khim.*, p. 743 (1964).
^{c}K. DeArmond, B. B. Garrett, and H. S. Gutowsky, *J. Chem. Phys.* **42**, 1019 (1965).
^{d}J. Owen and I. M. Ward, *Phys. Rev.* **102**, 591 (1956).
^{e}N. S. Garifyanov, B. M. Kozyrev, and V. N. Fedotov, *Dokl. Akad. Nauk SSSR* **156**, 641 (1964).
^{f}B. R. McGarvey, *Inorg. Chem.*, **5**, 467(1966).
^{g}R. H. Borcherts and C. Kikuchi, *J. Chem. Phys.* **40**, 2270 (1964).
^{h}P. G. Rasmussen, H. A. Kuska, and C. H. Brubaker, Jr., *Inorg. Chem.* **5**, 467 (1966).
^{i}W. H. From, P. B. Dorain, and D. R. Locker, *Bull. Am. Phys. Soc.* **11**, 220 (1966).

Ca_2PO_4Cl a value of $Q_{max} = +2.92$ and $\beta^2 = 0.77$ is obtained. This value of Q is about $+1$ greater than that found from the more complete analysis, and β^2 is about 0.1 lower.

Values of Q_{max} and β^2 for M^{5+} and M^{4+} ions obtained in this manner are given in Table V. For all the complexes listed in this table except $K_3Mo(CN)_8$ the ground-state orbital is d_{xy} and β^2 is a measure of π bonding to the ligands in the plane perpendicular to the main symmetry axis. Thus we can understand why K_2MoOF_5 has the largest value of Q_{max}, since the MoF bond should be the most ionic. The much smaller values for $K_2MoO(SCN)_5$ and $K_3Mo(CN)_8$ indicate extensive π bonding with the SCN^- and CN^- ligand ions.

ACKNOWLEDGMENTS

The support of the National Science Foundation is gratefully acknowledged.

REFERENCES

1. A. H. Maki and B. R. McGarvey, *J. Chem. Phys.* **29**, 31, 35 (1958).
2. D. Kivelson and R. Neiman, *J. Chem. Phys.* **35**, 149 (1961).
3. H. R. Gersmann and J. D. Swalen, *J. Chem. Phys.* **36**, 3221 (1962).
4. M. Sharnoff, *J. Chem. Phys.* **41**, 2203 (1964).
5. T. R. Reddy and R. Srinivasan, *J. Chem. Phys.* **43**, 1404 (1965).
6. D. Kivelson and S. Lee, *J. Chem. Phys.* **41**, 1896 (1964).
7. K. DeArmond, B. B. Garrett, and H. S. Gutowsky, *J. Chem. Phys.* **42**, 1019 (1965).
8. H. Kon and N. E. Sharpless, *J. Chem. Phys.* **42**, 906 (1965).
9. B. R. McGarvey, *J. Chem. Phys.* **38**, 388 (1963).
10. R. Lacroix and G. Emch, *Helv. Phys. Acta*, **35**, 592 (1962).
11. B. B. Garrett, K. DeArmond, and H. S. Gutowsky, *J. Chem. Phys.* **44**, 3393 (1966).
12. E. Banks, M. Greenblatt, and B. R. McGarvey, *J. Chem. Phys.* **47**, 3772 (1967).
13. A. Abragam, J. Horowitz, and M. H. L. Pryce, *Proc. Roy. Soc. (London)* **A230**, 169 (1955).
14. B. R. McGarvey, *J. Phys. Chem.* **71**, 51 (1967).
15. R. E. Watson, *Phys. Rev.* **119**, 1934 (1960).
16. R. S. Mulliken, *J. Chem. Phys.* **23**, 1833 (1955).
17. R. Lacroix, *Compt. Rend.* **252**, 1768 (1961).
18. H. H. Jaffe, *J. Chem. Phys.* **21**, 258 (1953).
19. H. H. Jaffe and G. O. Doak, *J. Chem. Phys.* **21**, 196 (1953).
20. E. Banks, M. Greenblatt, and S. Holt, to be published.
21. G. H. Azarbayejani and A. L. Merlo, *Phys. Rev.* **A137**, 489 (1965).
22. R. W. Kedzie, J. R. Shane, and M. Kestigian, *Phys. Letters (Netherlands)* **11**, 286 (1964).
23. B. R. McGarvey, *J. Chem. Phys.* **41**, 3743 (1964).

Magnetic Tensor Anisotropy in a Low-Symmetry Copper(II) Chelate*

Brian W. Moores and R. Linn Belford

Noyes Chemical Laboratory and Materials Research Laboratory
University of Illinois
Urbana, Illinois

Some anisotropic parameters of the spin Hamiltonian for copper doped into single crystals of orthorhombic bis-(N-methylsalicylaldiminato) nickel(II) are presented and discussed in terms of what can and what cannot be inferred from the directional properties of the g tensor and three hyperfine tensors. The most remarkable finding in this study is that the in-plane g anisotropy is unusually large, and the principal in-plane axes are oriented approximately along the metal–ligand bonds, unlike the optical absorption anisotropy and contrary to the previous findings for the g tensors of similar chelates.

INTRODUCTION

The study of the nature of metal–ligand bonding in transition-metal chelates is facilitated by the use of electron paramagnetic resonance (EPR). Following the pioneering work of Maki and McGarvey,[1] much use has been made of the technique by chemists. The parameters obtainable by EPR in which one is chiefly interested are the magnitudes and directions of such magnetic parameters as the g tensor and the hfs (hyperfine structure) tensors associated with the central metal and ligand atoms. If one then assumes a simple molecular orbital model for metal–ligand bonding, it is possible from these data to obtain estimates of atomic-orbital mixing coefficients and therefore arrive at a measure of metal–ligand bond covalency.

As a part of a study of the electronic structure of low-symmetry copper(II) chelates we have investigated the EPR spectrum of bis-(N-methyl-salicylaldiminato) copper(II) as a dilute substitutional impurity in the isomorphous single crystal of the diamagnetic nickel(II) chelate. A complete analysis of the anisotropic magnetic parameters permits an assignment of the electronic ground state and a consistent analysis of the covalencies of the five nondegenerate ligand-field molecular orbitals.[2] The data for this complex are more extensive than those ordinarily obtainable, so that more nearly unambiguous statements than usual ought to be possible. The purpose

*Supported by Advanced Research Projects Agency Contract SD-131 through the Materials Research Laboratory at the University of Illinois.

of this paper is to present the results of the determination of certain of the magnetic parameters of this complex, and to make some qualitative observations about the directions of their principal axes and what their directional properties imply (and do not imply) about the effective molecular symmetry.

EXPERIMENTAL

All EPR spectra were measured at room temperature using a standard Varian V4502-12 spectrometer with a 35-GHz (Q-band) microwave bridge and rotating magnet assembly, and employing 100-kHz magnetic field modulation. Single-crystal specimens were mounted on a quartz rod embedded in the tunable cavity and oriented by eye using a hand lens. This procedure, though crude, was satisfactorily reproducible. Magnetic-field intensities were read directly from a Varian Fieldial accessory, and a speck of powdered DPPH was used as an absolute external reference in the determination of g values. We took the isotropic g value of the free radical to be 2.0036.

Bis-(N-methylsalicylaldiminato) nickel(II) was prepared by the procedure of Klemm and Raddatz.[3] Doped samples containing about 0.1% ^{63}Cu were prepared by dissolving an appropriate quantity of ^{63}CuO in concentrated hydrochloric acid, evaporating the solution to dryness, and adding the brown anhydrous $CuCl_2$ to a hot saturated solution of the nickel chelate in absolute ethanol. Single crystals of the orthorhombic (α) modification were obtained by slowly cooling this solution in the presence of a boiling stone, which acted as a seed for the formation of needle-like red crystals. No satisfactory procedure was developed for the estimation of the absolute concentration of copper in the individual crystals; specimens of suitably low concentration were obtained by trial and error.

RESULTS AND INTERPRETATION

Crystallography

Bis-(N-methylsalicylaldiminato) nickel(II) occurs in at least two crystalline modifications, an orthorhombic and a monoclinic one, the crystal structures of both of which have been determined by three-dimensional x-ray techniques. The orthorhombic (α) form[4] is packed much like bis-(dimethylglyoximato) nickel(II),[5] and the planar molecules stack up in the unit cell with their molecular twofold (z) axes parallel to the [001] direction. The crystalline needle axis also lies in the [001] direction, and the well-developed face is (100). This fortunate crystal structure makes the α form ideally suited as a host for single-crystal EPR and optical studies, since the molecular axes are very simply related to any axis system conforming to the obvious external symmetry of the crystal. Figure 1 shows a view along [001] of two adjacent molecules in the unit cell. They are separated by 3.29 Å, and are related by a glide plane, as required in space group *Ibam*. With the static magnetic field along [100] or [010] their EPR spectra coalesce, and these

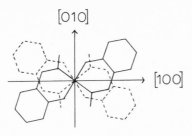

Fig. 1. View along [001] of two adjacent molecules in the orthorhombic modification of bis-(N-methylsalicylaldiminato) nickel(II).

coalescence points provide a useful experimental reference of magnetic field direction.

The crystal structure of the α form of the guest copper(II) chelate has also been determined;[6] Fig. 2 shows some relevant molecular parameters taken directly from that work, and also our choice of molecular axis system, with the molecular **z** axis taken perpendicular to the plane of the molecule.

EPR Spectra

Spectra were measured in a plane containing the **c** axis, and in the **ab** plane. In the former case the magnet angle was varied to maximize g, and this maximum g was taken to be equal to g_z. One of the four ^{63}Cu hfs components observed with the field in this direction is shown in Fig. 3. The observed 15-line pattern is consistent with a hyperfine interaction of the

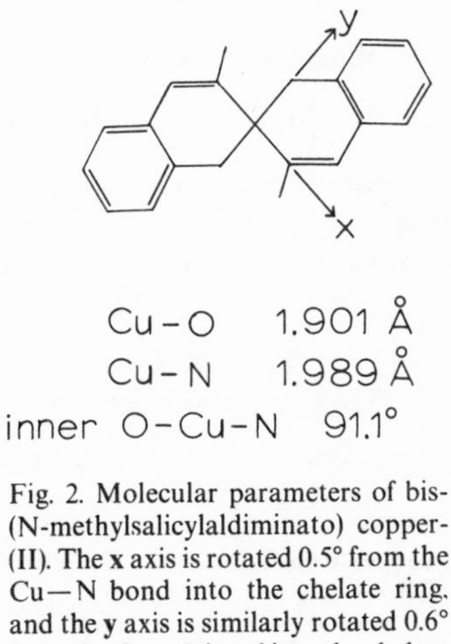

$$
\begin{array}{ll}
\text{Cu} - \text{O} & 1.901 \text{ Å} \\
\text{Cu} - \text{N} & 1.989 \text{ Å} \\
\text{inner O} - \text{Cu} - \text{N} & 91.1° \\
\end{array}
$$

Fig. 2. Molecular parameters of bis-(N-methylsalicylaldiminato) copper-(II). The x axis is rotated 0.5° from the Cu—N bond into the chelate ring, and the y axis is similarly rotated 0.6° from the Cu—O bond into the chelate ring.

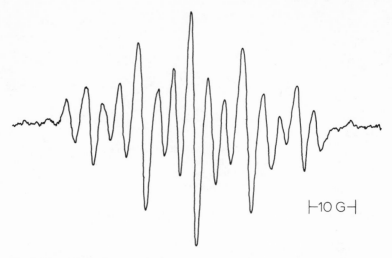

Fig. 3. Ligand hyperfine coupling structure on one of the four components of the
^{63}Cu hyperfine structure in orthorhombic bis-(N-methylsalicylaldiminato) copper(II).
The static magnetic field is along [001], which is the molecular z axis.

unpaired electron with two equivalent ^{14}N nuclei ($I = 1$) and two equivalent
protons ($I = \frac{1}{2}$).

For the measurements in the **ab** plane the g values and hyperfine
splittings for each of the two magnetically-inequivalent molecules in the
unit cell were measured at 15° intervals over a 180° rotation of the magnetic
field. The resulting data were fitted to the equation

$$g^2 = g_1^2 \cos^2\theta + g_2^2 \sin^2\theta$$

to provide the principal in-plane g values and to

$$K^2 g^2 = A_1^2 g_1^2 \cos^2\theta + A_2^2 g_2^2 \sin^2\theta$$

to provide the in-plane hfs components. The fitting process used a harmonic
analysis library subroutine on the University of Illinois IBM 7094 computer.
The angle θ is measured from the coalescence point along [010].

TABLE I
Magnetic Parameters

Parameter	Axis 1	Axis 2	Axis 3 ($= z$)
g	2.0561 ± 0.0007	2.0386	2.2157
$10^4 A^{(Cu)}$, cm^{-1}	(30.9 ± 0.2)	21.1	186.5
$10^4 A^{(N)}$, cm^{-1}	15.0	11.3	11.8
$10^4 A^{(H)}$, cm^{-1}	5.5	4.3	4.2

TABLE II

	g_2	$A_2^{(Cu)}$	$A_2^{(N)}$	$A_2^{(H)}$	x	y
θ, deg	39.5	34.3	28.5	24.9	−55	35

The relevant experimental results are given in Tables I and II. The θ values tabulated in Table II are those for the second principal parts of the magnetic parameters for only one of the two nonequivalent molecules, and the θ values for the x and y axes of that molecule are given for reference.

DISCUSSION

g Tensor

Table I shows that the directions of the principal axes of the g tensor differ only slightly ($\sim 4°$) from the metal–ligand bond axes. This result is not in accord with the findings of Maki and McGarvey,[1] who observed that in a similar complex, bis-(salicylaldiminato) copper(II), the principal in-plane g-tensor axes lie roughly along the bisectors of the inner and outer chelate angles, although the smallness of the in-plane anisotropy made the determination of these axes somewhat difficult. It is also at first sight surprising that the optical polarization directions for the N-methyl compound[7] also lie roughly along the bisectors of the chelate angles.

The apparent inconsistency of the EPR and optical results is understandable if we recognize[8,9] that the intensities and polarizations of optical bands in the centric Cu(II) planar complexes are determined by the effect of the admixture into the *gerade* upper state of the transition of certain *ungerade* low-energy ligand-to-metal charge-transfer states, accessible by strongly-allowed one-electron transfer from the ground state.[9] Consequently, the directional selection rules are determined by those charge-transfer bands. The directions of the principal axes of the g shift, however, are independent of these charge-transfer states, and are determined solely by the five *gerade* ligand field states of primarily metal d-orbital character. In general, therefore, and in particular for complexes of low symmetry,[9] there is no reason to suppose that the optical polarization directions need coincide with the g-shift principal axis directions.

In general, the symmetry and orientation of the g tensor will be related to the symmetry of the effective ligand field, and in *true* high-symmetry situations one can make unambiguous deductions about ligand-field directional properties from an analysis of the g tensor. In the chelate we consider here neither an appropriate set of pseudosymmetry axes for the effective ligand field in the xy plane nor an appropriate effective high-symmetry point group (other than the obvious C_{2h}) is readily apparent. In such a case the g-tensor orientation is determined by a combination of the geometry of the ground-state orbital $(x^2 - y^2)$, the extent to which it is contaminated with other d orbitals (xy, which provides a rotation, and $3z^2 - r^2$, which is introduced by a rhombic component of the ligand field oriented along the lobes of the effective ground-state orbital), and the orientation of the two excited-state orbitals derived from xz and yz. This situation is discussed by Hitchman *et al.*[10,11] We may safely assume that the orbital half-occupied in the ground state is essentially $x^2 - y^2$, with its four lobes pointing approximately toward the four ligand atoms. Then a very small admixture of $3z^2 - r^2$, which will extend two of the lobes (toward the nitrogens, say) at the expense of the other two, will tend to increase the difference between the two principal in-plane g values and to keep the principal axes of the g tensor directed toward the ligand. But the energies and orientation of the xz and yz orbitals are determined by the π system, and the effective π-system ligand-field rhombic component may be directed anywhere in the **xy** plane—along the bond axes, along the bisectors, or any-where in between. If xz is oriented towards the ligands, then one could describe the effective ligand field as being generally D_{2h}, the g principal axes would coincide with the bonds, and $g_x - g_y$ could be quite large. But if xz is directed so as to bond effectively with the π orbitals of both nitrogen and oxygen, the xz and yz orbitals will try to orient the g principal axes along the bond bisectors. The net result is a compromise, with the g principal axes oriented between the bonds and the bond-angle bisectors. The closer the g orientation is to the bisectors, the smaller is the effect of $3z^2 - r^2$ mixing on the in-plane g anisotropy. Thus, if one finds the g axes directed very nearly toward the bisectors, as is usually found for planar Cu(II) chelates, one can conclude that the effect of a ground-state rhombic field component *along* the bonds is probably negligible and that the excited-state $d\pi$ orbitals (xz and yz) are split and probably oriented along the bisectors. The usual expressions relating molecular-orbital parameters to the g shift then apply. But if one finds the g axes more nearly coincident with the bonds, then one does not know whether xz and yz are oriented in the same way or not, because the effects of $3z^2 - r^2$ mixing can dominate the effects of $d\pi$ aniso-tropy. Thus it is not safe to draw conclusions regarding the nature of the $d\pi$ orbitals from the usual expressions, which do not include the $d\pi - d_{z^2}$ competition.

It appears that for the system reported here the Zeeman and Cu hfs spin Hamiltonian is not quite rhombic, g being rotated by 6° from $A^{(Cu)}$, but is nearly so. The highest local symmetry consistent with the rhombic Hamiltonian would be D_{2h}. If it were a good approximation to neglect the $3z^2 - r^2$ component of the ground-state orbital $(x^2 - y^2)$, then the fact that the direction of g_2 deviates from the metal–oxygen vector by about 5° toward the interior of the chelate ring would imply that xz and yz are slightly mixed so as to rotate them half as much in the same direction. On the other hand, if there is a large effect from the $3z^2 - r^2$ component, then xz and yz will be mixed much more and effectively rotated farther than 5°. Calculations of covalency parameters based on each of these extremes of interpretation are given elsewhere.[2,12]

Metal Hyperfine Structure

Table II shows that the in-plane principal axes of the metal hfs tensor coincide within 0.1° (i.e., within the experimental uncertainty) with the metal–oxygen bond axis (and within 1.2° with the metal–nitrogen bond). This is to be expected if we suppose these axes to be determined primarily by the dipolar interaction of the nuclear spin with the spin of an electron in a molecular orbital composed primarily of the pure metal $d_{x^2-y^2}$ orbital, with the x- and y-directed lobes affected differently by $3z^2 - r^2$ contamination and by molecular orbital formation.

Ligand Hfs

The nitrogen in-plane hfs principal axes deviate by 6 to 7° from the metal–ligand bond axes. Figure 4 shows the directions of these axes. In interpreting this deviation, we note that the principal axes of the hfs tensor of any magnetic nucleus in the molecule are determined by the matrix elements of the anisotropic hfs operator averaged over the ground-state wave function. This wave function will contain a contribution from each atom in the molecule, with suitable weighting coefficients, such that any observed principal value of the hfs will approximately be the vector sum of contributions from unpaired electron density on each atom,[13] with positive

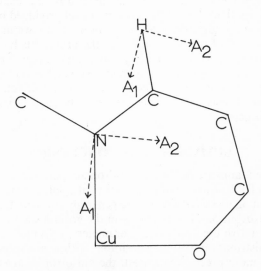

Fig. 4. Principal axis directions of in-plane ligand hfs in bis-(N-methylsalicylaldiminato) copper(II).

or negative contributions from overlap density in the bonding regions. In a simple, high-symmetry, paramagnetic salt such as $KNiF_3$ the ligand hfs principal axes will lie along, and perpendicular to, the metal–ligand bonds, since unpaired electron density is localized, to a good approximation, on the central metal ion and the first-coordination-sphere ligands. However, in a low-symmetry, covalent chelate such as the present one we may expect substantial asymmetries in the ligand hfs principal axes, since the unpaired electron density may be distributed over several atoms in a delocalized molecular orbital. Hence the observed hfs principal axes for the nitrogen nucleus in the present complex are consistent with a small unpaired electron density in the nitrogen–carbon bond, whose dipolar hfs interaction with the nitrogen nucleus would produce a slight rotation of the N hfs principal axes in a fashion shown in Fig. 4. If all the electron density were on the metal atom and the nitrogen and oxygen atoms, we should expect no such rotation, unless the Cu–N linkage involved a so-called bent bond. One must be wary of this interpretation, however, because the bulk of the N hfs is isotropic, and the anisotropic part and its orientation are subject to considerable error.

Similar arguments may be applied to the proton hfs, whose in-plane principal axes are also shown in Fig. 4. As shown by McConnell and Strathdee,[14] the principal axes of anisotropic proton hfs in a · CH fragment lie along, and perpendicular to, the C—H bond. In a paramagnetic radical of lower symmetry, however, or in one in which the proton is separated by several atoms from the principal paramagnetic center, the axes may depart significantly from the bond axes, as demonstrated experimentally by Atherton and Whiffen[15] in a study of a π radical produced by the γ-irradiation of single-crystal glycollic acid. In the present case we may interpret the observed directions of the components of the proton hfs by supposing that they are determined predominantly by concurrent unpaired electron density on the copper and nitrogen atoms, with some contribution from spin density on the adjacent carbon atoms. The magnitudes and directions of all three components are consistent with the spin-density distribution derived from the other parameters.[2,11,12]

SUMMARY AND CONCLUSIONS

1. Selected anisotropic spin-Hamiltonian parameters from a dilute single crystal EPR study of bis-(N-methylsalicylaldiminato) copper(II) in the corresponding nickel(II) chelate are generally consistent, and fit a partly-delocalized distribution of spin density in the ground state.

2. Quantitative use of all these parameters to fix most of the details of spin-density distribution would be desirable. Although the present study has perhaps been more precise than most, the anisotropies are nevertheless so small that some qualitative and quantitative uncertainties remain in their interpretation. Many questions could be answered by more precise experimental determinations of hfs parameters, and it may be that ENDOR

would be a useful experimental tool in sharpening our picture of the electronic ground state.*

3. The chelate, in the crystal form reported here, shows a strongly rhombic g anisotropy, which could be caused by a variety of factors. In particular, this anisotropy is not consistent with any *single* orientation of the xz and yz orbitals.

ACKNOWLEDGMENT

We wish to thank Dr. John J. Capart for useful discussions, and for performing many early exploratory measurements.

REFERENCES

1. A. H. Maki and B. R. McGarvey, *J. Chem. Phys.* **29**, 35 (1958).
2. B. W. Moores and R. L. Belford, to be published.
3. W. Klemm and K.-H. Raddatz, *Z. Anorg. Allgem. Chem.* **250**, 207 (1942).
4. M. R. Fox and E. C. Lingafelter, *Acta Cryst.* **22**, 943 (1967).
5. L. E. Godyicki and R. E. Rundle, *Acta Cryst.* **6**, 487 (1953).
6. E. C. Lingafelter, G. L. Simmons, B. Morosin, C. Scheringer, and C. Freiburg, *Acta Cryst.* **14**, 1222 (1961).
7. J. Ferguson, *J. Chem. Phys.* **35**, 1612 (1961).
8. R. Englman, *Mol. Phys.* **3**, 48 (1960).
9. R. L. Belford and J. W. Carmichael, *J. Chem. Phys.* **46**, 4515 (1967).
10. J. S. Griffith, *Phys. Rev.* **132**, 316 (1963).
11. M. A. Hitchman and R. L. Belford, this volume, Chapter 7.
12. B. W. Moores and R. L. Belford, Paper No. S106, Division of Physical Chemistry, American Chemical Society Meeting, San Francisco, California, April 2, 1968; also B. W. Moores, Ph.D. Thesis in Chemistry, University of Illinois, Urbana, Illinois, 1968.
13. W. Derbyshire, *Mol. Phys.* **5**, 225 (1962).
14. H. M. McConnell and J. Strathdee, *Mol. Phys.* **2**, 129 (1959).
15. N. M. Atherton and D. H. Whiffen, *Mol. Phys.* **3**, 1 (1960).

*Note added in proof: Hyde and co-workers have now successfully detected ENDOR in bis-(8-hydroxyquinolinate) copper (II). [G. H. Rist and J. S. Hyde, *J. Chem. Phys.* **49**, 2449 (1968)].

Ligand Hyperfine Couplings and the Structure of Transition-Metal Complexes*

Robert G. Hayes

Department of Chemistry
University of Notre Dame
Notre Dame, Indiana

Hyperfine couplings arising from nuclei in the ligands of transition metal complexes have been known for some fifteen years now and have from the first been valuable sources of information on covalency in such species. Within the past few years, however, many new results on ligand hyperfine couplings have been obtained, primarily by ESR. In particular, it has been shown in several systems that the often-invoked back donation or π covalency in cyanides is largely nonexistent, and the detection of ligand hyperfine couplings has verified, in several cases, an energy level scheme which puts the unpaired electron(s) in an orbital which is localized mainly on the ligands. A brief discussion of the detection and analysis of ligand hyperfine couplings will be followed by several examples of interest.

INTRODUCTION

Hyperfine structure arising from the nuclear spin in a ligand surrounding a transition-metal ion was first reported by Owen and Stevens.[1] They observed a splitting of the electron paramagnetic resonance of Ir^{3+} ions present in K_3InCl_6 as a dilute impurity, caused by interaction of the unpaired electrons with nuclei in the surrounding Cl^-. Stevens was able to calculate the extent of delocalization of the unpaired electron onto the chlorides from the observed splitting.

Many observations of ligand hyperfine couplings have been reported since that of Owen and Stevens. Most of the observations have been extra splittings of the electron paramagnetic resonance, but several ligand hyperfine couplings have been observed by using nuclear magnetic resonance. In nuclear magnetic resonance ligand hyperfine couplings appear not as a splitting of a resonance, but as a shift of the resonance from the ligand nuclei. This shift is proportional to the ligand hyperfine coupling and also to the magnetization of the electrons. If the shift is observed as a function of the direction of the magnetic field in a single crystal, one can obtain the extent of delocalization of the unpaired electron, or electrons, onto the ligands from the data. The classic work in the observation of ligand hyperfine

*Supported by National Science Foundation through grant NSF-GP-7881.

couplings by NMR is that of Shulman and Jaccarino,[2] who studied several fluorides.

A third technique which has been used occasionally to study ligand hyperfine couplings is electron nuclear double resonance, or ENDOR. In this technique the electron paramagnetic resonance is partially saturated and a radio-frequency signal is applied to the sample. The extent of saturation of the electron paramagnetic resonance changes when the frequency of the rf signal corresponds to one of the spacings of the nuclear magnetic levels. One is thus able to measure the splittings of nuclear levels through the EPR. Kravitz and Piper[3] discuss the application of ENDOR to the measurement of ligand hyperfine splittings in FeF_6^{3-}. ENDOR will probably be used more in the future because it permits a more precise measurement of ligand hyperfine couplings and because the ENDOR spectrum of a system in which several nuclei are coupled to the electrons is less complex than the EPR spectrum. The number of lines in an ENDOR spectrum increases arithmetically with the number of nonequivalent nuclei, but the number of lines in the EPR spectrum increases geometrically with the number of nonequivalent nuclei. This advantage is counterbalanced by the greater difficulty of the ENDOR experiment.

We shall present an outline of the methods used in analyzing ligand hyperfine splittings observed in EPR, then discuss a few of the observations. We shall not attempt a complete survey of the ligand hyperfine couplings which have been reported. Many reports have appeared. A simple list of observations would be of little value, and an attempt to analyze each report would occupy too much space. A review article on covalency in transition-metal ions has appeared recently.[4] This review article discusses observations on fluorides at length and discusses some of the observations on other halides briefly.

ANALYSIS OF LIGAND HYPERFINE SPLITTINGS

As we have mentioned, ligand hyperfine couplings appear as extra splittings in EPR spectra. One accounts for them by adding terms to the spin Hamiltonian representing couplings between the electron magnetic moment of the system and the various ligand nuclear magnetic moments. One must, then, also include terms representing various other interactions between the ligand nuclei and their surroundings, such as quadrupole couplings and the direct nuclear Zeeman interaction. The assumption is made that the couplings between the electronic system and the several nuclei may be treated independently, so the spin Hamiltonian takes the form

$$\mathscr{H} = \mathbf{H} \cdot \mathbf{g} \cdot \mathbf{S} + \mathbf{I} \cdot \mathbf{A}_c \cdot \mathbf{S} + \cdots + \sum_i \mathbf{I}_i \cdot \mathbf{A}_{L,i} \cdot \mathbf{S}$$

$$+ \sum_i \mathbf{I}_i \cdot \mathbf{Q}_i \cdot \mathbf{I}_i + \sum_i g_{N,i} \beta_N \mathbf{H} \cdot \mathbf{I}_i \qquad (1)$$

The first two terms represent typical terms in the usual spin Hamiltonian, and the last three, containing sums over the various ligand nuclei, represent the ligand hyperfine coupling, the ligand nuclear quadrupole coupling, and the direct ligand nuclear Zeeman coupling, respectively.

The various tensors appearing in Eq. (1) may all have different principal axes, a point worth remembering. In a common case the g and A_c tensors have axial symmetry, and ligand hyperfine couplings due to equatorial ligands are observed. The ligand hyperfine couplings often have nearly axial symmetry, but the unique axis is *not* the unique axis of the g and A_c tensors but, rather, the metal–ligand bond direction. It is not sufficient, then, to measure ligand hyperfine splittings with the magnetic field along the unique axis of g and perpendicular to this axis.

The analysis of observed hyperfine splittings to obtain the parameters of Eq. (1) is often complicated because the several terms involving ligand nuclear spins have similar magnitudes. The selection rule $\Delta m_I = 0$ is meaningless, and each nucleus contributes as many as $(2I + 1)^2$ transitions of unequal intensity and spacing. If there are several nuclei, the spectrum becomes very complex.[5] One usually seeks angles where all nuclei are equivalent, or only a few nuclei have substantial hyperfine couplings, and so on, depending on the system. The assignment is usually verified by computer simulation of the spectrum at various angles.

The ligand hyperfine tensors which are obtained from the data contain contributions from the electron spin and orbital magnetic moment distributions of the entire complex. One wishes to separate out the contribution from electrons on the ligand in question. Marshall[6] has described the corrections in great detail. They arise from spin and orbital magnetic moment distributions in the rest of the molecule, and from orbital angular momentum on the ligand. In the simplest case, which is often encountered, one has a single orbital state. This means that the ground state is orbitally nondegenerate and spin-orbit coupling does not mix orbital states very much. Then the correction is only for spin magnetic moment on the central metal ion. This can be approximated by the magnetic moment of a point dipole centered on the central metal nucleus. The hyperfine tensor of such a point dipole has axial symmetry about the metal–ligand bond direction, with principal values $2A_D$, $-A_D$, $-A_D$,

$$A_D = \tfrac{2}{5}g\beta g_N\beta_N R^{-3} \tag{2}$$

where R is the metal–ligand bond length. This correction is typically about 1 MHz for ^{14}N, and about 10 MHz for ^{19}F. More sophisticated corrections have been required in some cases (see, e.g., Thornley *et al.*[7]).

The corrected ligand hyperfine coupling tensor is supposed to arise from unpaired electron density in ligand orbitals. Suppose that after correction the ligand hyperfine tensor has principal values A, B, C. Decompose the tensor into an isotropic part,

$$\mathbf{A} = \tfrac{1}{3}(A + B + C) \tag{3}$$

and a traceless part, a, b, c, defined by

$$a = A - \mathbf{A} \tag{4}$$

with similar expressions for b and c. Then one has

$$
\begin{aligned}
f_S &= 2SA/A_S \\
a &= (A_p/2S)(2f_A - f_B - f_C) \\
b &= (A_p/2S)(2f_B - f_A - f_C) \\
c &= (A_p/2S)(f_C - f_A - f_B)
\end{aligned}
\tag{5}
$$

where A_s is the hyperfine coupling which would arise from an unpaired electron in a ligand valence-shell s orbital, which is given by

$$A_s = (8\pi/3)g_n\beta_n g_0\beta|\psi(0)|^2_{vs} \tag{6}$$

and A_p is half the coupling constant in the direction of the orbital which would arise from an electron in a valence-shell p orbital,

$$A_p = \tfrac{2}{5}g_n\beta_n g\beta\langle r^{-3}\rangle_{vp} \tag{7}$$

Usually, A_s and A_p are gotten from Hartree–Fock calculations, although they may be gotten from atomic data. Morton[8] cites two recent tabulations, and a set calculated from Clementi's functions[10] appears in Table I. The values of A_s and A_p gotten from different sources vary by 10–15%.

TABLE I
Values of A_s and A_p for the Ground States
of Some Neutral Atoms

Atom	A_s, GHz	A_p, MHz
^{13}C	3.11	90.5
^{14}N	1.54	47.8
^{19}F	47.9	1514.
^{31}P	9.88	238.
^{35}Cl	4.66	123.
^{75}As	9.50	278.

The equations for a, b, and c are not independent, because $a + b + c = 0$. Only two of f_A, f_B, and f_C, or two linear combinations, can be obtained. If the hyperfine tensor has axial symmetry, there is another restriction, that two of the f's be equal, and only one quantity, $f_\sigma - f_\pi$, can be obtained.

Further analysis of the ligand hyperfine couplings is based, as a rule, on molecular orbital theory. The linear combinations of ligand orbitals which can bond to the several d orbitals, in octahedral symmetry, are well known.

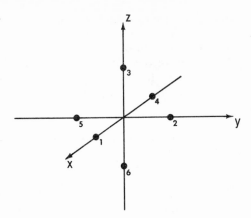

Fig. 1. Coordinate system for an octahedral complex.

If one uses the coordinate system of Fig. 1, they are

$$\chi_s(x^2 - y^2) = \tfrac{1}{2}(s_1 - s_2 + s_4 - s_5)$$

$$\chi_s(z^2) = (1/2\sqrt{2})(2s_3 + 2s_6 - s_1 - s_2 - s_4 - s_5)$$

$$\chi_p(x^2 - y^2) = \tfrac{1}{2}(-p_{x,1} + p_{y,2} + p_{x,4} - p_{y,5})$$

$$\chi_p(z^2) = (1/2\sqrt{2})(-2p_{z,3} + 2p_{z,6} + p_{x,1} + p_{y,2} - p_{x,4} - p_{y,5}) \qquad (8)$$

$$\chi_p(xy) = \tfrac{1}{2}(p_{y,1} + p_{x,2} - p_{y,4} - p_{x,5})$$

$$\chi_p(xz) = \tfrac{1}{2}(p_{z,1} + p_{x,3} - p_{z,4} - p_{x,6})$$

$$\chi_p(yz) = \tfrac{1}{2}(p_{z,2} + p_{y,3} - p_{z,5} - p_{y,6})$$

These expressions ignore ligand overlap in the normalization. Expressions for symmetry orbitals in other geometries can be found in various places. Owen and Thornley[4] have tabulated ligand symmetry orbitals in several geometries. The tables may be obtained from them.

Ligand symmetry orbitals may be combined with central metal orbitals to form bonding and antibonding molecular orbitals. Taking $d_{x^2-y^2}$ as an example, and considering only bonding to s orbitals for simplicity, one has

$$\psi_B = N_B[\chi_s(x^2 - y^2) + \gamma\, d_{x^2-y^2}]$$
$$\psi_A = N_A[d_{x^2-y^2} - \lambda\chi_s(x^2 - y^2)] \qquad (9)$$

One can show from orthonormality that

$$\lambda = (\gamma + S)/(1 + \gamma S) \qquad (10)$$

where S is the overlap integral, $\langle d_{x^2-y^2}|\chi(x^2 - y^2)\rangle$. One thus sees that λ is not zero even if the bonding orbital is localized entirely on the ligands, so that

γ is zero. This means that, strictly speaking, one should not call $N_A\lambda$ the covalency, but rather $N_B\gamma$. It also means something more significant. The central metal orbitals must be orthogonalized to all the ligand orbitals.[11] Thus in the simple case above one must orthogonalize the $d_{x^2-y^2}$ to the inner s orbitals of the ligands. If we call $\chi_{1s}(d_{x^2-y^2})$ the symmetry orbital based on $1s$ orbitals, and let $\langle d_{x^2-y^2}|\chi_{1s}(d_{x^2-y^2})\rangle = S_1$, we have

$$\psi_A = N_A[d_{x^2-y^2} - S_1\chi_{1s}(x^2 - y^2) - \lambda\chi_s(x^2 - y^2)] \qquad (11)$$

If the unpaired electron density at a ligand nucleus be evaluated on this function, one gets, in addition to the term $N_A^2\lambda^2|\psi(0)|_{vs}^2$, several terms, of which the most important is $N_A^2\lambda S_1\langle\psi_{1s}(0)\psi_{vs}(0)\rangle$. One might be inclined to ignore all such terms on the argument that S_1 is very small, but $|\psi_0(0)|_{1s}^2$ is large. It has been shown that about half the ligand hyperfine couplings in transition-metal fluorides can be ascribed to simple orthogonalization.[11]

We have estimated the integrals necessary for such an analysis for CuPc, in which the ^{14}N hyperfine couplings are known.[12]

We used Clementi's wave functions[10] and took the Cu—N bond distance to be 2.48 Å. The integrals appear in Table II. The observed hyper-

TABLE II
Analysis of ^{14}N Hyperfine Data of CuPc

$$\chi_1 = \tfrac{1}{2}[(1s)_1 - (1s)_2 + (1s)_3 - (1s)_4]$$

$$\chi_2 = \tfrac{1}{2}[(2s)_1 - (2s)_2 + (2s)_3 - (2s)_4]$$

$$\langle\chi_1|d_{x^2-y^2}\rangle = 0.2270 \qquad \langle\chi_2|d_{x^2-y^2}\rangle = 0.0052$$

$$|\psi(0)|_{1s}^2 = 98.19 \text{ A.U.} \qquad |\psi(0)|_{2s}^2 = 4.77 \text{ A.U.} \qquad |\psi_{1s}(0)\psi_{2s}(0)| = 21.64 \text{ A.U.}$$

$$A_{1s} = +31{,}600 \text{ MHz} \qquad A_{2s} = +1533 \text{ MHz} \qquad A_{1s,2s} = -6956 \text{ MHz}$$

$$\phi = N(d_{x^2-y^2} - \langle\chi_1|d_{x^2-y^2}\rangle\chi_1 - \lambda\chi_2)$$

$$|\phi(0)|^2 \doteq \tfrac{1}{4}N^2\lambda^2|\psi(0)|_{2s}^2 + \tfrac{1}{2}N^2\langle\chi_1|d_{x^2-y^2}\rangle\lambda|\psi_{1s}(0)\psi_{2s}(0)|$$

$$\lambda = 0.3762 \qquad \gamma = 0.163$$

fine coupling is nearly isotropic, with value 48.8 MHz. An elementary analysis based on Eq. (9) yields $N_A\lambda 0.357$, from which one obtains $\gamma = 0.123$. If one includes the $1s$ orbital in the analysis, one finds that the term $N_A^2S_1\lambda$ contributes about -6.5 MHz, so the value of $N_A\lambda$ rises to 0.381. We thus see that the $1s$ orbitals on nitrogen make a substantial contribution to the hyperfine coupling, about 12%.

SOME EXPERIMENTAL RESULTS

We shall restrict ourselves to two examples from the large number of studies which has been reported.

Table III displays the data which are available on cyanides. Three points are to be observed. First, the iron-series compounds have little π covalency.

TABLE III
Ligand Hyperfine Couplings in Cyanide-Containing Species

	f_s, %	f_p, %	Ref.
$Cr(CN)_6^{3-}$	-2.20	~ 0	b
$Fe(CN)_6^{3-}$	-0.88	1.0–3.0	c
$Cr(CN)_5NO^{3-}$			
Equatorial	$(-)0.92$	3.0	d
Axial	$(-)0.62$	0.	
$Mo(CN)_5NO^{3-(a)}$	(-0.88)	$(+5.3)$	e
$Mo(CN)_8^{3-}$	$(-)0.83$	—	f
$Fe(CN)_5NOH^{2-}$	$(+)0.75$	—	g

$^a f_s$ calculated from $A_\parallel$; may be in error by 20%. f_p refers to ^{14}N.
[b] H. A. Kuska and M. T. Rogers, *J. Chem. Phys.* **41**, 3802 (1964).
[c] D. G. Davis and R. J. Kurland, *J. Chem. Phys.* **46**, 388 (1967).
[d] J. J. Fortman and R. G. Hayes, *J. Chem. Phys.* **43**, 15 (1965).
[e] R. G. Hayes, *J. Chem. Phys.* **47**, 1692 (1967).
[f] S. I. Weissman and M. Cohn, *J. Chem. Phys.* **27**, 1440 (1957).
[g] J. Danon, R. P. A. Muniz, and A. O. Caride, *J. Chem. Phys.* **46**, 1210 (1967).

Three systems have been studied: $Cr(CN)_6^{3-}$ is t_{2g}^3, $^4A_{2g}$; $Fe(CN)_6^{3-}$ is t_{2g}^5, $^2T_{2g}$; and $Cr(CN)_5NO^{-3}$, which has a strong axial distortion, is $e_g^4 b_2$, $^2B_{2g}$. As Table III shows, the fraction of unpaired electron in π orbitals on carbon is less than 0.03 in all cases. In some cases there are conflicting interpretations of the amount of covalency.

The ^{14}N hyperfine couplings are not yet available for any iron-series cyanide complex. It is difficult to decide just how large the nitrogen hyperfine couplings could be without resolved hyperfine lines being observed. Judging from the experience gained from the study of $Mo(CN)_5NO^{3-}$, which has resolved ^{14}N splittings (see Table III), one would guess that the spin density on ^{14}N in $Cr(CN)_5NO^{-3}$ could not be much more than 0.01.

In contradistinction to the iron-series compounds, $Mo(CN)_5NO^{3-}$ has an easily-resolved ^{14}N hyperfine splitting from the equatorial cyanides, from which one obtains a π spin density of 0.053 per cyanide. Unfortunately, ^{13}C hyperfine data for $Mo(CN)_5NO^{3-}$ are very fragmentary, so one cannot complete the argument in this case either.

The isotropic splittings of all cyanide-containing complexes with unpaired electrons in metal $d\pi$ orbitals show a curious regularity, as Table III shows. In a wide variety of cases in which an unpaired electron is an orbital of π type with respect to a cyanide the isotropic coupling constant corresponds to about 0.009 electron. This regularity is observed not only in compounds of similar geometry, but extends to $Mo(CN)_8^{3-}$, which has Archimedean antiprismatic geometry.[13] One can extend the rule to $Cr(CN)_6^{3-}$. In this case there are two electrons in π-type orbitals and one in a $d\,\delta$ orbital. Taking values from the axial and equatorial cyanides in $Cr(CN)_5NO^{3-}$, one comes very close to the observed spin density.

The invariance of the isotropic coupling constant to ^{13}C in the species discussed is difficult to rationalize. There can be no direct bonding in these cases, of course. It has been customary to assume that spin polarization of σ-bonding orbitals by the unpaired electron causes the spin density in carbon s orbitals. The implications of this proposal have been worked out in some detail for $Cr(CN)_5NO^{3-}$.[14]

As reference to Danon $et\ al.$[14] will show, a similarity of ^{13}C isotropic coupling constants in compounds as varied as $Cr(CN)_6^{3-}$, $Cr(CN)_5NO^{3-}$, $Mo(CN)_5NO^{3-}$, and $Mo(CN)_8^{3-}$ is not to be expected. The coupling constants depend on the details of the bonding and on the distribution of excited states. These are surely not the same in all compounds studied.

We can offer no rationalization of the observations. If the mechanism discussed by Danon $et\ al.$[14] is correct, there must be an unexpected cancellation between various parameters. Even if some other mechanism were important, such as polarization of the cyanide ion by the adjacent unpaired spin, it is difficult to believe that the extent of polarization would be the same in all cases.

Finally, we wish to draw attention to the ^{13}C coupling in $Fe(CN)_5NOH^{2-}$, in which one unpaired electron resides in d_{z^2}. It should be mentioned that some debate exists about the assignment of a spectrum to this species and about the electronic structure of the species.[15,16] We subscribe to the conclusions of van Voorst and Hemmerich.[15] The odd thing about the ^{13}C coupling constant is its smallness. We assume it to be positive. One obtains an s spin density of 0.0075, which is very small. If one assumes sp hybridization, the occupancy of the sp hybrid is then 0.015. This is not much for a ligand, which is supposed to form σ bonds. The spin polarization, of opposite sign, presumably persists, and might double the value estimated.

In summary, studies of ^{13}C hyperfine couplings in cyanides have raised more questions than they have answered.

The role of π covalency has been elucidated, but the role of spin polarization is far from clear. In the one observed case the σ covalency is much less than one might have expected. One needs some theoretical work on the spin polarization, as well as some more experiments on systems with nuclei other than carbon. A few data are available on ^{14}N couplings. These are tabulated in Table IV. Once again, the isotropic coupling constants in several compounds with unpaired electron in an orbital of π type with respect to the nitrogen are nearly the same.*

Some more data on ^{13}C couplings in compounds with unpaired electrons in σ orbitals would be very useful.

We conclude with one more example, which is more easily interpreted. Table V shows the Cl^- hyperfine couplings observed in $CuCl_6^{4-}$ and $AgCl_6^{4-}$. It is seen that the covalency involves p orbitals on Cl^- almost exclusively. Table V also carries out the analysis, which yields the expected result that Ag^{2+} forms more covalent complexes than does Cu^{2+}.

*The nominal ground states of the VO^{2+} complexes are $^2B_{2g}(d_{xy})$. $Cr(bipy)_3^+$ has a $^2A_1(d_{z^2})$ ground state, quantizing along the trigonal axis.[17]

TABLE IV
Some ^{14}N Hyperfine Couplings

	f_s, %	Ref.
Cr(dipy)$_3^+$	0.56	a
V(dipy)$_3^0$	0.42	a
VO(porphyrin)	0.50	b

[a]E. König, Z. Naturforsch. **19**, 1139 (1964).
[b]D. Kivelson and S. K. Lee, J. Chem. Phys. **41**, 1896 (1964).

TABLE V
Ligand Hyperfine Analysis of Two d^9 Chlorides$^{(a)}$

	A, MHz	B, MHz	f_s, %	f_p, %	Ref.
CdCl$_2$:Cu^{2+}	55.5	15.0	0.64	7.82	b
KCl:Ag^{2+}	97.4	15.0	0.95	15.90	c

[a]Cl$^-$ parameter from Ref. b.
[b]J. H. M. Thornley et al. Proc. Phys. Soc. **78**, 1263 (1961).
[c]J. Sierro, J. Phys. Chem. Solids **28**, 417 (1967).

It is clear that observation of ligand hyperfine couplings permits one to make precise estimates of covalency. Unfortunately, in most cases one can only obtain some of the information about covalency from ligand hyperfine couplings, and the rest must be gotten by indirect methods. The study of ligand hyperfine couplings has also raised some interesting questions. The questions are hard to answer, though, because the large number of electrons in these systems makes detailed calculations difficult.

REFERENCES

1. J. Owen and K. W. H. Stevens, Nature **171**, 836 (1953).
2. R. G. Shulman and V. Jaccarino, Phys. Rev. **42**, 666 (1956).
3. L. C. Kravitz and W. W. Piper, Phys. Rev. **146**, 322 (1966).
4. J. Owen and J. H. M. Thornley, Rep. Prog. Phys. (London) **29**, 675 (1966).
5. T. P. P. Hall, W. Hayes, R. W. H. Stevenson, and J. Wilkens, J. Chem. Phys. **38**, 1977 (1963); H. M. Gladney, Phys. Rev. **143**, 198 (1966); R. G. Hayes, J. Chem. Phys. **47**, 1692 (1967).
6. W. Marshall, in: Paramagnetic Resonance, Vol. 1, W. Low, ed. (Academic Press, New York, 1963).
7. J. H. M. Thornley, C. G. Windor, and J. Owen, Proc. Roy. Soc. **248A**, 252 (1965).
8. J. R. Morton, Chem. Rev. **64**, 453 (1964).
9. C. M. Hurd and P. Coodin, J. Phys. Chem. Solids **28**, 523 (1967).
10. E. Clementi, IBM J. Res. Dev. **9**, 2 (1965).
11. A. J. Freeman and R. E. Watson, Phys. Rev. Lett. **6**, 343 (1961).

12. S. E. Harrison and J. M. Assour, in: *Paramagnetic Resonance*, Vol. 2, W. Low, ed. (Academic Press, New York, 1963).
13. B. R. McGarvey, *Inorg. Chem.* **5**, 476 (1966).
14. J. Danon, H. Panepucci, and A. A. Misetich, *J. Chem. Phys.* **44**, 4154 (1966).
15. J. D. W. van Voorst and P. Hemmerich, *J. Chem. Phys.* **45**, 3914 (1966).
16. J. Danon, R. P. A. Muniz, and O. A. Caride, *J. Chem. Phys.* **46**, 1210 (1967).
17. L. Garber, Department of Chemistry, Notre Dame University, unpublished observations.

Computer Synthesis of Electron Paramagnetic Resonance Spectra from a Parametric (Spin) Hamiltonian*

John H. Mackey, Marcel Kopp, Edmund C. Tynan, and Teh Fu Yen

Mellon Institute
Carnegie–Mellon University
Pittsburgh, Pennsylvania

A computer program has been developed for computing electron paramagnetic resonance (EPR) spectra from a spin Hamiltonian, $\mathcal{H}_{sp}$, which is a linear combination of spin operators whose coefficients may be regarded as experimental parameters. The user supplies $\mathcal{H}_{sp}$, the spin number (orientation degeneracy) of each particle, and values of the parameters; output may be obtained in the form of the *frequency* spectrum for fixed applied magnetic-field strength, H_0, or the magnetic-field spectrum for fixed radiation frequency, ν_{mr}. In addition to the line positions, the program computes first-order transition intensities in a radiation field H_1 using eigenvectors generated by the calculation. Results have been obtained for (transition-metal) ions and free radicals in single crystals as a function of field orientation, and in powders and glasses (by summing over many orientations). The program can be adapted for calculation of NMR and NQR spectra of anisotropic systems, inclusion of line narrowing and broadening (relaxation) effects, and automatic adjustment of certain parameters to fit experimental spectra.

INTRODUCTION

The *spin Hamiltonian*, $\mathcal{H}_{sp}$, plays a key role in the analysis of experimental paramagnetic resonance spectra. For a paramagnetic system in a magnetic field H_0, $\mathcal{H}_{sp}$ is an approximate energy operator with stationary states connected by transitions of low (microwave) frequency. There exists the formal relation: $\mathcal{H}_m \leftrightarrow \mathcal{H}_{sp} \leftrightarrow$ Experimental spectrum. It is usually impractical to proceed directly from the molecular Hamiltonian ($\mathcal{H}_m$) to the experimental spectrum, since $\mathcal{H}_m$ contains both spatial and spin operators; its solution involves molecular orbitals, spacial integrals, etc. However, for very general interactions among particles and with applied fields one can write an equivalent energy function $\mathcal{H}_{sp}$ containing only spin operators. Its

*Sponsored by the Air Force Office of Scientific Research, Arlington, Virginia, under Grant AFOSR 974–66/67. Additional support was received from the Gulf Research & Development Company as part of the research program of the Multiple Fellowship on Petroleum.

equivalence to $\mathscr{H}_m$ (within a group of low-lying states) is expressed in the coefficients of the various spin operators constituting $\mathscr{H}_{sp}$.

In practice, the spectra may be analyzed (in simpler situations) to provide the parameters for an $\mathscr{H}_{sp}$ of the appropriate form, or spectra may be synthesized by inserting in $\mathscr{H}_{sp}$ a set of approximate parameters and finding the eigenfunctions and eigenvalues by standard quantum-mechanical methods. The program, MAGNSPEC,* takes the assumed parameters and uses matrix methods to generate a synthetic spectrum by scanning the applied magnetic field systematically. A time-dependent (transition-inducing) interaction with a monochromatic H_1-field is included to first order using semiclassical perturbation theory.

THEORETICAL BASIS

Synthesis of the paramagnetic resonance spectrum amounts to the solution for several fixed values of H_0 and H_1 of the time-dependent Schrödinger equation

$$\mathscr{H}_{sp}|\psi(t)\rangle = -i\hbar\frac{\partial}{\partial t}|\psi(t)\rangle \tag{1}$$

Suppose that $\mathscr{H}_{sp}$ can be written as the sum of a large, time-independent term, $\mathscr{H}_{stat}$, and a small, time-dependent term, $\mathscr{H}_{rad}$. One first solves the static problem

$$\mathscr{H}_{stat}|v_k\rangle = E_k|v_k\rangle \tag{2}$$

then $\mathscr{H}_{rad}$ is included and (1) is solved in the approximation that $|\psi(t)\rangle$ is a linear combination of the stationary eigenstates, $|v_k\rangle$, with time-dependent coefficients, so that the populations of states $|v_k\rangle$ are linear functions of the duration of application of $\mathscr{H}_{rad}$ (see, e.g., Schiff[2]).

As usual,[2] suppose that $|v_k\rangle$ is expanded in terms of a complete set of orthonormal states (basis vectors) $|u_m(j)\rangle$, which are eigenfunctions of J^2 [eigenvalue $= j(j + 1)$] and $J_z (ev = m)$, J being some angular momentum operator. Thus†

$$|v_k\rangle = \sum_m |u_m(j)\rangle\langle u_m|v_k\rangle \tag{3}$$

where $\langle u_m|v_k\rangle$ is a scalar product (vector component). If $\mathscr{H}$ is some energy function which can be written in terms of angular momentum operators,

$$\mathscr{H}|v_k\rangle = |v_l\rangle = \sum_{m,m'} |u_{m'}(j)\rangle\langle u_{m'}|\mathscr{H}|u_m\rangle\langle u_m|v_k\rangle \tag{4}$$

---

*A copy of the FORTRAN deck and a user's manual have been sent to the Quantum Chemistry Program Exchange, Indiana University; also see Kopp and Mackey.[1]

†This is analogous to the expansion of a vector $\mathbf{v}$ in ordinary three-space in a basis $\mathbf{e}_1, \mathbf{e}_2, \mathbf{e}_3$ as $\mathbf{v} = \mathbf{e}_1(\mathbf{e}_1 \cdot \mathbf{v}) + \mathbf{e}_2(\mathbf{e}_2 \cdot \mathbf{v}) + \mathbf{e}_3(\mathbf{e}_3 \cdot \mathbf{v})$; however, the scalar products $\langle u_m|v_k\rangle$ are taken in a complex Hilbert n-space.

where

$$\langle u_{m'}|\mathscr{H}|u_m\rangle = \mathscr{H}_{m'm} \equiv \int u_m^*(j)\mathscr{H}u_m(j)\,d\tau \tag{5}$$

is a matrix element of $\mathscr{H}$ in the $|u_m\rangle$ representation. Now, $|v_l\rangle$ in (4) is another vector in the space spanned by $|u_m\rangle$, so

$$|v_l\rangle = \sum_{m'}|u_{m'}(j)\rangle\langle u_{m'}|v_l\rangle \tag{6}$$

Using (4) and (6), the transformation $|v_l\rangle = \mathscr{H}|v_k\rangle$ may be written in the equivalent form

$$\langle u_{m'}|v_l\rangle = \sum_m \langle u_{m'}|\mathscr{H}|u_m\rangle\langle u_m|v_k\rangle \tag{7}$$

The coefficients $\langle u_m|v_k\rangle$ $(m = j, j-1, \ldots, -j)$ may be regarded as the components of a column vector; in particular, the basis vectors themselves are simply the set

$$\begin{bmatrix} \delta_{j,m} \\ \delta_{j-1,m} \\ \vdots \\ \delta_{-j,m} \end{bmatrix}$$

In this notation the eigenvalue problem (2) is written

$$\sum_m \langle u_{m'}|\mathscr{H}_{\text{stat}}|u_m\rangle\langle u_m|v_k\rangle = E_k\langle u_{m'}|v_k\rangle \tag{8}$$

Consider the set of eigenvectors $\langle u_m|v_k\rangle$ for all eigenvalues E_k. These are orthogonal, i.e.,

$$\sum_m \langle v_l|u_m\rangle\langle u_m|v_k\rangle = \delta_{lk} \tag{9}$$

hence, they may be treated as the columns of a unitary matrix, U_{mk}, and (8) may be written

$$\sum_m \mathscr{H}_{m'm}U_{mk} = E_k U_{m'k} \tag{10}$$

Using (9),

$$\sum_{m',m} U_{lm'}^+\mathscr{H}_{m'm}U_{mk} = \sum_{m'} U_{lm'}U_{m'k}E_k = \delta_{lk}E_k = D_{lk} \tag{11}$$

or

$$U^+ \cdot \mathscr{H}_{\text{stat}} \cdot U = D \tag{11'}$$

with $D = [D_{lk}]$, a diagonal matrix.

Thus the solution of (2) reduces to finding a unitary transformation U (element $U_{mk} = \langle u_m/v_k\rangle$), which reduces the hermitian matrix $\mathscr{H}_{\text{stat}}$(element $\mathscr{H}_{mm'}$), to diagonal form D (element $E_k\delta_{lk}$).

Returning to Eq. (1), let

$$|\psi(t)\rangle = \sum_k a_k(t)|v_k\rangle$$

Then (1) becomes

$$\sum_k a_k(t)\langle v_j|\mathscr{H}_{rad}(t)|v_k\rangle = -ih\sum_k \dot{a}_k(t)\langle v_j|v_k\rangle = -ih\dot{a}_j(t) \qquad (12)$$

Let us seek a solution to (12) with $\mathscr{H}_{rad}(t) = \mathscr{H}_{rad}(0)e^{i\omega t}$ and $a_j(t = 0) = 1$, $a_k(0) = 0$ $(k \neq j)$. The corresponding form of (12),

$$\dot{a}_j(t) = -ih\sum_k a_k(t)e^{i\omega t}\langle v_j|\mathscr{H}_{rad}(0)|v_k\rangle \qquad (13)$$

may be solved by standard methods[2] to give

$$I_{jk} \propto |\langle v_j|\mathscr{H}_{rad}(0)|v_k\rangle|^2 \qquad (14)$$

where I_{jk} is the intensity of the transition from state k to state j or the power absorbed in the resonance line $j \leftarrow k$.

To reduce (14) to matrix-vector notation, we use (3) to get

$$\langle v_j|\mathscr{H}_{rad}|v_k\rangle = \sum_{m',m} \langle v_j|u_{m'}\rangle\langle u_{m'}|\mathscr{H}_{rad}(0)|u_m\rangle\langle u_m|v_k\rangle$$

$$= \sum_{m',m} U^+_{jm'}\mathscr{H}_{rad}(0)_{m'm}U_{mk}$$

$$= (U^+\mathscr{H}_{rad}(0)U)_{jk}$$

so

$$I_{jk} \propto |(U^+\mathscr{H}_{rad}(0)U)_{jk}|^2 \qquad (15)$$

FLOW DIAGRAM OF MAGNSPEC

A schematic flow diagram for the main program is shown in Fig. 1. It shows the sequence of reading the data and calling subroutines. The basic program logic is as follows:

Boxes 1–8. Read the data and set up the matrix of $\mathscr{H}_{stat}$ at the largest field to be considered (or, alternately, at a selected test field). Overall job control is exercised by INDEX (column 1 of the first data card) which is zero (or blank) for a new job, nine (9) for termination of computation, and 1–5 for the two-parameter interpolation procedure TETRAFIT (see next section). The field orientation is read at boxes 5 and 6; LAST ($= 99$) is the termination for a group of calculations at several field orientations. HSPINEST establishes the matrix $\mathscr{H}_{stat}$ at zero field, and HSPINMOD appends the field-dependent terms.

Boxes 9–11. Diagonalize HSP(J, K) at $H_0 = H_{max}$ (or $H_0 = H_{test}$) and generate the frequency spectrum and intensities (using the matrix of $\mathscr{H}_{rad}$) at this field value (box 10). For lines MALLD(J,K) of appreciable intensity

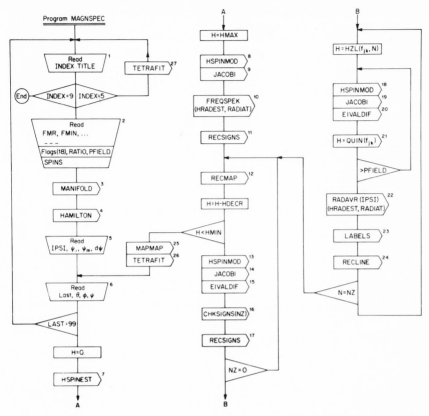

Fig. 1. Flow diagram for MAGNSPEC showing the order of reading data and calling subroutines by the main program.

record the signs (box 11) of the differences between the transition frequency, FALLD(J,K) and the radiation frequency, FMR.

Boxes 12–24. Adjust HSP for the next field, $H - H_{\text{decr}}$, prediagonalize with approximate eigenvectors (those obtained from the last diagonalization at H), complete the diagonalization, and compare signs of FALLD(J,K) − FMR with those at the last field (box 16). If any sign changes have occurred (NZ > 0), obtain the resonant field, HZ(J,K), at which FALLD − FMR = 0, by interpolation (boxes 18–21) to desired precision (PFIELD) and determine (box 22) the line intensity, RADINT(J,K). Record HZ(J, K), RADINT(J,K), and (J,K). The loop 18–24 is executed N times until all transitions in the current field interval have been tracked to resonance (N = NZ). Control is then returned to Box 12, where the positions, intensities, and labels are recorded (RECMAP). The field is stepped and the whole process 13–24 repeated.

Box 25 to End. When H_{min} has been passed control is transferred to MAPMAP, which uses the storage of RECMAP, etc., for mapping of the magnetic spectrum. If the spectrum is desired for a new field orientation, control is transferred from Box 5 to Box 7. When the next orientation card has LAST = 99, control is returned to Box 1. When the next title card has INDEX (column 1) = 9, the computation is finished.

PROGRAM DETAILS

Setting Up the Spin Representation

Subroutine MANIFOLD (box 3) generates real matrix representations of the spin components of the particles and stores them in an array, $S(J,K,N)$. If ND1 is the number of particles, then $\Pi_{I=1}^{ND1} [2 \, SPIN(I) + 1] = ND2$ is the dimension of these matrices; ND2 is also the number of basis vectors needed to span the spin space. The matrices S_z, S_+ ($= S_x + iS_y$), and $S_-(= S_x - iS_y)$ are computed in the uncoupled-particle or direct-product representation (e.g., Corio[3]). These direct products are obtained from the spin matrices within the subspace of particle i whose nonzero elements are

$$\langle u(j_i,m)|S_{zi}|u(j_i,m)\rangle = m$$

$$\langle u(j_i,m \pm 1)|S_{\pm i}|u(j_i,m)\rangle = [j_i(j_i + 1) - m(m \pm 1)]^{1/2} \qquad (16)$$

$$\langle u(j_i,m)|S_{zk}|u(j_i,m)\rangle = \langle u(j_i,m)|S_{\pm k}|u(j_i,m)\rangle = 1$$

$$(i \neq k)$$

The results are stored in successive sheets of $S(J,K,N)$; $N = 1$, ND1.

Specifying the Spin Hamiltonian Form

Function HAMILTON specifies the detailed algebraic forms of $\mathscr{H}_{stat}$ and $\mathscr{H}_{rad}$, whose matrix elements are stored in the arrays $HSP(J,K)$ and $HRAD(J,K)$, respectively. This function has four entry points: HAMILTON, HSPINOP, ZEEMAN, and HRADOP. At the first call, HAMILTON (box 4) reads values of the parameters in the particular form chosen by the user, and performs any preliminary computation. Later calls by HSPINEST (to the other entry points) generate elements (J,K) of HSP or HRAD one at a time.

For the one- and two-particle systems studied so far, two types of FUNCTION HAMILTON have been used; these may be denoted as HAMILTON (SCALAR) and HAMILTON (TENSOR). In HAMILTON (SCALAR) the elements of HSP and HRAD are specified as explicit linear combinations of the components of the spin vectors, their products, and cross products. The coefficients in these scalar forms contain angular variables (θ, ϕ, ψ), which specify the orientations (Fig. 3, lower) of the fields H_0 and H_1 with respect to a set of references axes in the crystal or molecular system under study; they also contain the parameters $(g_x, g_y$, etc.)

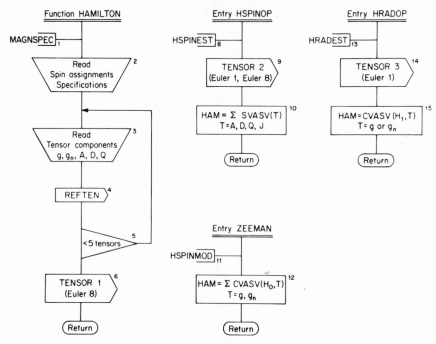

Fig. 2. Flow chart and entry points to the function subroutine HAMILTON (TENSOR).

read by HAMILTON. The user is free to specify $\mathcal{H}_{stat}$ and $\mathcal{H}_{rad}$ forms and quantization axes in any way provided that he includes algorithms for selecting the proper elements of $S(J,K,N)$ and computing the elements of $HSP(J,K)$ and $HRAD(J,K)$.

To relieve the user from the burden of writing a scalar HAMILTON function for each new type of $\mathcal{H}_{stat}$, a more general routine using tensor forms is supplied. This HAMILTON (TENSOR) assumes a quadratic static Hamiltonian of the form[4,5]*

$$\mathcal{H}_{stat} = (\beta_e \tilde{H}_0 \cdot g_e \cdot S) + (\beta_n \tilde{H}_0 \cdot g_n \cdot I) + (\tilde{S} \cdot A \cdot I)$$
$$+ (\tilde{S} \cdot D \cdot S) + (\tilde{I} \cdot Q \cdot I) + (J\tilde{S} \cdot S) \qquad (17)$$

It includes (in order) an electron Zeeman term, a nuclear Zeeman term, an electron–nuclear hyperfine term, a quadratic electron spin–spin term, a nuclear quadrupole term, and a (scalar) exchange coupling.

It is assumed that each of the second-order tensors g_e, g_n, A, D, and Q is symmetrical. By two successive orthogonal transformations (rotations), the tensor components are transformed from the coordinate systems in which they are diagonal to a coordinate system defined by the fields H_0, H_1. Finally, each spin is quantized along H_0, which is the z axis of the final frame.

*More recently this tensor form has been extended to three nuclei.

A choice of radiation interactions of the form

$$\mathscr{H}_{rad} = \beta_e \tilde{H}_1 \cdot \mathbf{g}_e \cdot \mathbf{S} \tag{18}$$

or

$$\mathscr{H}_{rad} = \beta_n \tilde{H}_1 \cdot \mathbf{g}_n \cdot \mathbf{I} \tag{19}$$

is provided for use in conjunction with $\mathscr{H}_{stat}$ specified in (17).

The program flow of HAMILTON (TENSOR) is shown in Fig. 2. At the first call, the proper assignment of the spins to the operators S, I occurring in (17) are read-in (box 2). Also (box 3), the characteristic roots (g_{xx}, g_{yy}, g_{zz}), and the directions $(\theta_{xx}, \phi_{xx}, \text{etc.})$ of the principal axes of g in a convenient reference frame, **abc**, are read (see Fig. 3, upper). REFTEN

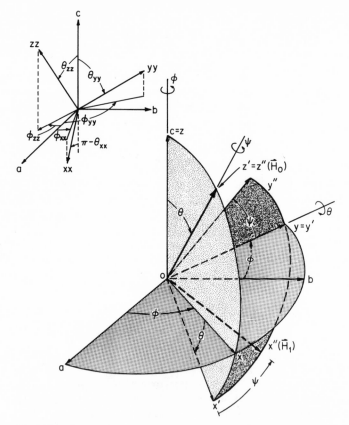

Fig. 3. Coordinate systems used for the spin-Hamiltonian representations. (*Upper*) Polar angles θ_{xx}, ϕ_{xx}, etc., of the principal axis vectors (xx, yy, zz) with respect to the crystal or molecular reference basis (a, b, c). (*Lower*) Euler angles ϕ, θ, ψ which rotate the reference basis into coincidence with the field basis $[x''(\mathbf{H}_1), y'', z''(\mathbf{H}_0)]$.

(box 4) converts the polar angles to direction cosines in the reference frame, checks that the vectors are accurately orthogonal, and forms the rotation matrix, TRG. TRG transforms components from the reference frame to the g frame:

$$TRG = \begin{bmatrix} \sin\theta_{xx}\cos\phi_{xx} & \sin\theta_{xx}\sin\phi_{xx} & \cos\theta_{xx} \\ \sin\theta_{yy}\cos\phi_{yy} & \sin\theta_{yy}\sin\phi_{yy} & \cos\theta_{yy} \\ \sin\theta_{zz}\cos\phi_{zz} & \sin\theta_{zz}\sin\phi_{zz} & \cos\theta_{zz} \end{bmatrix} \quad (20)$$

If g_d and g_{ref} denote the forms of the g tensor in the diagonal (principal axis) and reference frames, respectively, then

$$g_d = (TRG)\cdot g_{ref}\cdot(\widetilde{TRG}) \quad (21)$$

where $\sim$ designates transposition. Since TRG is orthogonal,

$$g_{ref} = \widetilde{TRG}\cdot g_g\cdot TRG \quad (22)$$

In a similar manner (box 5), the components of all the interactions in Eq. (17) are read and converted to matrices TRGN, etc., of the form (20).

The tensors are transformed to the reference frame by the call TENSOR $(1,\ldots)$ (box 6), which specifies a transformation of the form (22); the matrix triple product is actually executed by EULER $(8,\ldots)$. When return is made to the main program, the *reference frame* forms of the tensors are left in storage.

At later entries from HSPINEST (box 8), etc., the tensors are transformed to the field-based coordinate axes $(x'' = H_1, y'', z'' = H_0)$. This transformation (Fig. 3, lower) may be regarded as a product of three elementary rotations involving the Euler angles ϕ, θ, and ψ;* hence the matrix TRF (from reference frame to field frame) is

$$TRF = T(\psi)T(\theta)T(\phi) \quad (23)$$

and the transformation of the components of a tensor g_{ref} in the reference frame to the field frame g_{field} is simply

$$g_{field} = TRF\cdot g_{ref}\cdot\widetilde{TRF} \quad (24)$$

This operation is prescribed by TENSOR $(2,\ldots)$ (box 9) and carried out by EULER $(7,\ldots)$. The scalar components are then evaluated by routines SVASV (box 10) and CVASV (box 12 and 15) using the spin assignments specified. Later entries to ZEEMAN (box 11) and HRADOP (box 13) result in similar operations.

Matrix Diagonalization

The eigenvalue problem, e.g., (11), is solved by a generalization to hermitian matrices[5,6] of the Jacobi method of diagonalizing real, symmetrical

---

*The specification of the third angle, ψ, is not necessary to the calculation of the correct eigenvalues; the value chosen for ψ does generally affect the intensities [call TENSOR $(3,\ldots)$].

matrices.[6,7] It consists in reducing off-diagonal elements to zero systematically by successive applications of elementary transformations R_i involving a single row and column. Given a real symmetrical matrix A, the set A_i, where $A_1 = R_1^{-1} A R_1$, $A_2 = R_2^{-1} A_1 R_2$, etc. converges to diagonal form to any desired precision. At the same time the set $U_1 = R_1$, $U_2 = R_1 \cdot R_2$, etc. approaches the (orthogonal) eigenvector matrix U (but not to the same precision).

We wish to diagonalize the $n \times n$ *hermitian* matrix $A + iB$, i.e., to solve the eigenvalue problem

$$(A + iB)(V + iW) = \Lambda(V + iW) \tag{25}$$

For any column, i.e., for any eigenvector $(v + iw)$,

$$(av - bw) + i(aw + bv) = \lambda(v + iw) \tag{26}$$

Using the $2n \times 2n$ symmetrical matrix

$$\left[\begin{array}{c|c} A & B \\ \hline -B & A \end{array} \right]$$

the problem may be expressed as

$$\left[\begin{array}{c|c} A & B \\ \hline -B & A \end{array} \right] \left[\begin{array}{c|c} V & W \\ \hline -W & V \end{array} \right] = \left[\begin{array}{c|c} \Lambda & 0 \\ \hline 0 & \Lambda \end{array} \right] \left[\begin{array}{c|c} V & W \\ \hline -W & V \end{array} \right] \tag{27}$$

The standard Jacobi procedure, which may be applied to (27), generates the solution to (25) in duplicate, as may be seen by expanding (27) and comparing column by column with (26).

Other matrix transformations are also carried out with the expanded matrices; e.g., the intensity coefficients, Eq. (15) were generated by

$$\left[\begin{array}{c|c} \tilde{V} & \tilde{W} \\ \hline -\tilde{W} & \tilde{V} \end{array} \right] \left[\begin{array}{c|c} H_{\mathrm{rad}}^r & H_{\mathrm{rad}}^i \\ \hline -H_{\mathrm{rad}}^i & H_{\mathrm{rad}}^r \end{array} \right] \left[\begin{array}{c|c} V & W \\ \hline -W & V \end{array} \right]$$

$$= \left[\begin{array}{c|c} (U^+ H_{\mathrm{rad}} U)^r & -(U^+ H_{\mathrm{rad}} U)^i \\ \hline -(U^+ H_{\mathrm{rad}} U)^i & (U^+ H_{\mathrm{rad}} U)^r \end{array} \right] \tag{28}$$

which is fully equivalent to (in duplicate)

$$[\tilde{V} - i\tilde{W}][H_{\mathrm{rad}}^r + iH_{\mathrm{rad}}^i][V + iW] = [(U^+ H_{\mathrm{rad}} U)^r + i(U^+ H_{\mathrm{rad}} U)^i] \tag{29}$$

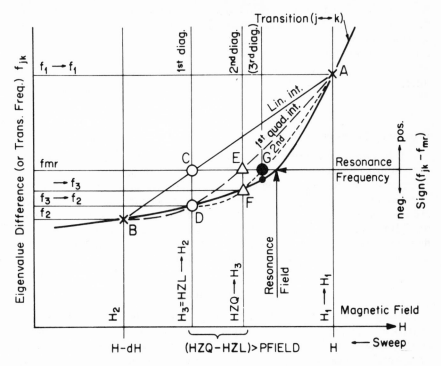

Fig. 4. Schematic plot of transition frequency vs magnetic field for a transition $j \to k$. The actual variation is shown by the solid curve ($BDFA$). Points C, E, and G are successive two- and three-point interpolations of the resonance field for the fixed radiation frequency f_{mr}.

Field Interpolation

Details of the field-scanning and interpolation procedures are shown in Fig. 4. If transition $j \to k$ (solid curve $AFDB$) crosses resonance ($f_{jk} - f_{mr}$ changes sign) in the interval $H \to H - dH$, a linear interpolation between A and B is obtained at HZL(C). $\mathcal{H}_{stat}$ is diagonalized at HZL to yield a third point (D) on the transition-energy curve. Using the three points (A, B, and D), a parabolic interpolation to the resonant frequency is obtained at HZQ (E). If |HZQ − HZL| > PFIELD, the desired field precision, $\mathcal{H}_{stat}$ is rediagonalized at HZQ to yield point F and a second quadratic interpolation (to G on DFGA) is taken. This loop may be reexecuted until PFIELD is secured. Often the field vs frequency curves are nearly linear, and the values of HZL and HZQ are already the same to a precision PFIELD.

Interpolation of $\mathcal{H}_{STAT}$ Parameters

The routine TETRAFIT (Fig. 1) has been included as a simple method of simultaneously adjusting any two parameters, p and q, in $\mathcal{H}_{stat}$ on the basis of coincidence of the computed and experimental positions of lines of the spectrum. The ranges of p and q [(p_1, p_2) and (q_1, q_2), respectively] are set, and four spectra for the pairs (p_1, q_1), (p_1, q_2), (p_2, q_1), and (p_2, q_2) are calculated with INDEX = 1, 2, 3, 4 in sequence. On a fifth pass with INDEX = 5 experimental positions of lines are read-in, and the double-linear interpolations of p and q fitting each line are given as output. This procedure applies to single orientations only and has some advantage over an intuitive approach.

Program Options and Controls

It has been said earlier that the program allows the choices: frequency-spectrum/field-spectrum and NMR-spectrum/ESR-spectrum. There are additional options for parameter units, radiation-field orientation, manner of labeling transitions, and diagonalization procedure. In addition, there are controls governing field precision (PFIELD), diagonalization accuracy (RATIO), minimum resonance-line intensity (RALLD), and a series of 18 flags, most of which control print out. These options and controls enable the user to limit the detail during execution to fit his needs.*

APPLICATIONS

The program has been applied mostly to one- or two-spin paramagnetic centers in single crystals, polycrystalline samples, and glasses. Since MAGN-SPEC calculates only zero-width spectra for single orientations, an auxiliary program,[5] (SPREAD),† is used to generate plots of line-broadened spectra and powder and glass spectra.

One-Spin Systems

Fine Structure of Cr^{3+}

The results of computation may be illustrated for the simple one-particle system ^{52}Cr ($I = 0$) lightly-doped into α-Al_2O_3, pink ruby. In tensor form the static Hamiltonian is

$$\mathscr{H}_{stat} = (\beta_e \tilde{H}_0 \cdot g_e \cdot S) + (\tilde{S} \cdot D \cdot S) \tag{30}$$

with $S = \frac{3}{2}$; the environment of Cr^{+3} (trigonally-distorted octahedral) requires that $\mathscr{H}_{stat}$ have axial symmetry.[9] If we choose the crystal axes as the reference basis (a,b,c), then c is the axis of distortion. The tensors g_e and D are diagonal in this coordinate system, so that $\mathscr{H}_{stat}$ can be expanded to

---

*See first footnote p. 34.

†The program SPREAD processes the output of MAGNSPEC, i.e., it takes as input the δ-function spectra (positions, intensities, and labels) produced by MAGNSPEC for one or more field orientations, assigns a shape and width to each line, and accumulates the spectra. SPREAD can also generate additional spectra by interpolation between the computed orientations, and thus produce a facsimile of a powder pattern. Finally, it can renormalize and superimpose several envelopes, each with a different statistical weight.

give

$$\mathscr{H}_{stat}(sym) = \beta H_0[g_{\parallel}(\cos \theta)S_z + \tfrac{1}{2}g_{\perp}(\sin \theta)(S_+ e^{-i\phi} + S_- e^{i\phi})]$$

$$+ D[S_z^2 - \tfrac{1}{3}S(S + 1)] \qquad (31)$$

where $g_{cc} = g_{\parallel}$, $g_{aa} = g_{bb} = g_{\perp}$, $D_{cc} = \tfrac{2}{3}D$, $D_{aa} = D_{bb} = -\tfrac{1}{3}D$, and θ is the angle between $\mathbf{H}_0$ and c.

On the other hand, if we expand (30) in a coordinate system (x',y',z'), with $z' \parallel \mathbf{H}_0$, and quantize S along the applied field, then

$$\mathscr{H}_{stat}(field) = \beta H_0[\tfrac{1}{2} (g_{\perp} - g_{\parallel})(\sin \theta)(\cos \theta)(S_+ + S_-)$$

$$+ (g_{\perp} \sin^2\theta + g_{\parallel} \cos^2\theta)S_{z'}]$$

$$+ D(\cos^2\theta - \tfrac{1}{2}\sin^2\theta)[S_{z'}^2 - \tfrac{1}{3}S(S + 1)]$$

$$+ \tfrac{1}{4}\sin^2\theta(S_+^2 + S_-^2)$$

$$- \tfrac{1}{2}(\sin \theta)(\cos \theta)(S_+ S_{z'}$$

$$+ S_{z'}S_+ + S_- S_{z'} + S_{z'}S_-) \qquad (32)$$

in terms of the same parameters.

Earlier calculations on ruby had been made by E. O. Schulz-du Bois,[9] who used the field-based Hamiltonian (32). The eigenvalues and eigenvectors were reported in the range of fields 0 and 10 kOe, for θ at intervals of 10°. In addition, magnetic-field spectra were given for the resonant frequencies 5.18, 6.08, 9.30, 12.33, 18.2, and 23.9 GHz. Schulz-du Bois used the parameters $D = -0.19155 \, \text{cm}^{-1}$, $g_{\parallel} = 1.9840$, and $g_{\perp} = 1.9867$, and found a satisfactory fit to the experimental line positions at 9.30 and 12.33 GHz.

To test our program, we calculated the frequency spectra at regular field intervals and the magnetic-field spectra at 9.30 GHz, using both scalar forms (31) and (32) for $\mathscr{H}_{stat}$, with the Schulz-du Bois parameters. The eigenvalues and eigenvectors (except for phase) agree in detail with his earlier results. Later calculations using the HAMILTON (TENSOR) routine were in complete agreement.

As a more definitive check against experimental data, measurements were made on a lightly-doped ruby sample* at 77°K and a radiation frequency of 9.5405 GHz. These were compared with the magnetic-field spectrum computed at this frequency. Initially, with Schulz-du Bois' parameters the fit at $\theta = 0$ and 90° was not satisfactory. Better values were obtained by systematically interpolating calculated line positions at these orientations (as will be described below) (values indicated by asterisks in Fig. 5); the values $D = -0.19110 \, \text{cm}^{-1}$, $g_{\parallel} = 1.9822$, and $g_{\perp} = 1.9841$ gave an adequate fit.

The results of calculations using the latter parameters are plotted in Fig. 5. Computed intensities at 10° intervals are indicated along the curves.

---

*This sample was kindly supplied to the authors by Dr. D. R. Feldman of the Westinghouse Research Center, Pittsburgh, Pennsylvania.

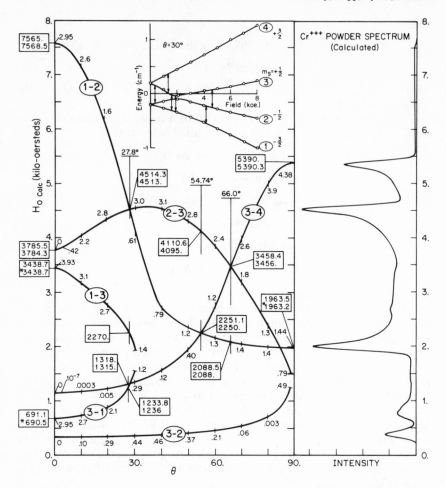

Fig. 5. Computed resonance fields and intensities as a function of the angle θ that the applied field makes to the main symmetry axis for Cr^{3+} in α-Al_2O_3 (pink ruby). The transitions are specified by the eigenvalue labels as indicated in the energy-level diagram shown in the inset (upper center) for $\theta = 30°$. The irradiation frequency was 9.540$_5$ GHz. The powder envelope synthesized from the calculated positions is shown at the right.

The only program difficulty was experienced at the loop closures near 30° and 90°: Since the transitions on opposite sides of the loop have a level in common,* both resonances could be missed if they fall within a single field-scanning interval (HDECR); the problem is circumvented by reducing HDECR in these regions.

Since the lines traverse very large field ranges as orientation is varied, it was not practical to set an arbitrary angle experimentally with any accept-

---

*The labels for the levels are in order of increasing energy (see inset to Fig. 5).

able accuracy. Therefore additional measurements were made at those orientations (Fig. 5) at which two lines cross; this criterion eliminates the need for very precise angle measurements. Experimental and calculated results are compared in the blocks in Figure 5 (measured resonance field on top).

At $\theta = 0°$ three lines (1–2, 1–3, and 3–1) are allowed and have both predicted and observed intensities in the ratio 3:4:3; the other three transitions should have zero intensity. The experimental results show that transition 2–3, which has an appreciable intensity of 0.42 at 2°, can actually be observed at 0° as a weak line; this is evidence of some mosaic structure within the sample. Transition 3–4, which is weak for $\theta < 25°$ (10^{-7} at 2°), is unobservable. The sixth transition (3–2) is at such low fields that it could not be identified at any orientation chosen for study.

Thus at $\theta = 0°$ there are four lines which may be used to set the two parameters $g_{\parallel}$ and D. Since the position of line 1–3 ($\frac{1}{2} \to -\frac{1}{2}$) does not involve D, this line can be used to set $g_{\parallel}$ uniquely; the agreement between the calculated and measured positions is then exact. One of the other three positions (which depend on both $g_{\parallel}$ and D) may be used to set D. The line 1–2 at high fields was very asymmetrical (apparently from distortions of the lattice) and its exact position was uncertain. Assuming that the line at low fields (3–1) was most independent of $g_{\parallel}$, we interpolated two calculations linearly to get D; there was a residual error of 0.6 Oe in the calculated position.

At $\theta = 90°$ there are two lines (3–4 and 1–2). Again the line at higher fields was highly asymmetrical; therefore the line at lower fields (1–2) was used to get $g_{\perp}$. The misfit was 0.3 Oe.

Experimental and calculated positions at 0, 27.8, 54.7 ($\cos^2\theta = \frac{1}{3}$), 66.0, and 90° are given in Fig. 5; in general, the positions are within a few oersteds of those calculated. Intensities were not measured accurately in view of the rather large changes in width and shape of the lines; qualitatively, they agree with those calculated.

Because of its interest in connection with powder and glass spectra discussed later, we have generated the powder pattern drawn along the right side of Fig. 5 using the second program (SPREAD). MAGNSPEC calculations were done at intervals of 5° (19 calculations). A three-point interpolation of position and intensity was used to subdivide the interval four times* (for a total of 289 orientations), and each line was broadened to a Gaussian shape with a width of 50 Oe before accumulating the results with a statistical weight proportional to $\sin\theta$. The experimental powder spectrum has not been reported.

Nuclear Quadrupole Resonance of ^{59}Co

Single-particle systems involving a nucleus with $I > \frac{1}{2}$ have $\mathcal{H}_{\text{stat}}$ analogous to (30), namely

$$\mathcal{H}_{\text{stat}} = (\beta_n \tilde{H}_0 \cdot g_n \cdot I) + (\tilde{I} \cdot Q \cdot I) \qquad (33)$$

where Q is the nuclear quadrupole interaction tensor. MAGNSPEC has been

*See first footnote p. 34.

used[10] to calculate such spectra for single-crystal $K_3Co(CN)_6$ involving ^{59}Co $(I = \frac{7}{2})$. The results are in very good agreement with experimental data and compare favorably with the best perturbation calculations.

Two-Spin Systems

Three cases of two-spin systems are described below. These range in type from calculations of single-crystal spectra in which the $\mathscr{H}_{stat}$ is rather complex (case a), to a powder-type spectrum in which $\mathscr{H}_{stat}$ is axial and of rather simpler form (case b), to a glass pattern which requires not only powder averaging, but a rather large statistical distribution of one of the parameters (case c). These examples are chosen from the direct research interests of the authors.

Case a. An Aluminum-Related Hole Center in α-Quartz

This center was discovered by Griffiths, Owen, and Ward (GOW)[11] and a model was proposed by O'Brien and Pryce.[12,13] This model is in qualitative accord with EPR spectra at 4.2, 20, and 77°K. It is used here to make our calculations fit our experimental spectra at 77°K for two special orientations ($H_0 \parallel$ c axis and $H_0 \parallel$ quadrupole axis). In tensor form the spin Hamiltonian is of the type

$$\mathscr{H}_{stat} = (\beta_e H_0 \cdot g \cdot S) - (\beta_n H_0 \cdot g_n \cdot I) + (I \cdot A \cdot S) + (I \cdot Q \cdot I) \quad (34)$$

with $S = (-)\frac{1}{2}$, $I = \frac{5}{2}$ (^{27}Al). In the O'Brien model the hole is predominantly on an oxygen which bridges a normal lattice silicon to a substitutional aluminum. This center has six positions in the unit cell, each with its g tensor axial along Si—Al and its hyperfine tensor and internal quadrupole field axial along Al—O. Initially, we used the GOW parameters shown in Table I,

TABLE I
Tensor Principal Axes in Reference Frame

	XX	θ	ϕ	YY	θ	ϕ	ZZ	θ	ϕ
G	2.00	90.0	90.0	2.00	32.0	180.0	2.06	58.0	0
GN	1.455400	90.0	0	1.455400	90.0	90.0	1.45540(	0	0
A	0.00056	90.0	60.0	0.00056	45.0	150.0	0.000480	45.0	30.0
D	-0	-0	-0	-0	-0	-0	-0	-0	-0
Q	-0.0000133	90.0	60.0	-0.0000133	45.0	150.0	$+0.0000267$	45.0	-30.0
J	-0	—	—	—	—	—	—	—	—

where the reference system is chosen (Fig. 3) with c parallel to the threefold axis and a parallel to one twofold axis of the α-quartz structure. The agreement between computed and measured spectra for $H_0 \parallel$ c (lines of all six centers coincide by symmetry) and for $H_0 \parallel A_{zz}, Q_{zz}$ (lines with $\Delta m_I \neq 0$ very weak) was not very good.

To obtain the fit shown in Fig. 6, we calculated the tensor directions specified by the O'Brien model from the structure data of Smith and

Alexander.[14] By trial and error the values of $g_\parallel$ and $g_\perp$ were adjusted to 2.0590 and 2.0045, respectively, and the angle $\theta(g_{zz})$ to a value (58.90°), close to that specified by GOW. This caused good registration of the groups as a whole, but left the hyperfine splittings and quadrupole coupling (which influences the forbidden line intensities) in poor correspondence. The spectrum for $H_0 \parallel A_{zz}, Q_{zz}$ was easily fitted, but several more adjustments were needed to improve the fit for $H_0 \parallel c$. The final parameters used are shown in Table II.

TABLE II
Tensor Principal Axes in Reference Frame

	XX	θ	ϕ	YY	θ	ϕ	ZZ	θ	ϕ
G	2.004500	90.0	84.2	2.004500	31.1	174.2	2.059000	58.9	−5.8
GN	1.455400	90.0	0	1.455400	90.0	90.0	1.455400	0	0
A	0.000590	90.0	64.1	0.000590	45.8	154.1	0.000465	44.2	−25.9
D	−0	−0	−0	−0	−0	−0	−0	−0	−0
Q	−0.000018	90.0	64.1	−0.000018	45.8	154.1	+0.000036	44.2	−25.9
J	−0	—	—	—	—	—	—	—	—

The calculated and experimental spectra are compared in Fig. 6. In practice, the stick diagrams (Fig. 6, center) were used to adjust the parameters. The top spectrum was synthesized from the "experimental" stick diagram, and the bottom spectrum was synthesized from the last computed stick diagram, in each case using a Lorentzian line shape with a width of 1.0 Oe. Considering the complexity of the c-axis spectrum and the number of parameters, it does not appear reasonable to attempt a better fit on the basis of just two orientations.

Case b. VO^{2+} *Etio porphyrin II in Castor Oil*

In this and the next example the calculations are based on the Hamiltonian

$$\mathcal{H}_{\text{stat}} = (\beta_e H_0 \cdot g \cdot S) + (I \cdot A \cdot S) \tag{35}$$

in which g and A are diagonal in the same coordinate system. To expedite the computation, $\mathcal{H}_{\text{stat}}$ was expressed explicitly in the so-called *effective field basis*;* this results in an $\mathcal{H}_{\text{stat}}$ representation that is diagonal in the Zeeman term and diagonal to first order in the hyperfine term, and so reduces the computation time as much as possible. In this quantization system

$$\mathcal{H}_{\text{stat}} = gH_0 S_{z'} + A(S'_+ I''_+ + S'_- I''_-) + B(S'_- I''_+ + S'_+ I''_-)$$
$$+ CS'_z I''_z + D(S'_- I''_- - S'_+ I''_+ + S'_- I''_+ - S'_+ I''_-)$$
$$+ E(S'_+ + S'_-)I''_z + F(S'_- - S'_+)I''_z \tag{36}$$

*The pertinent transformations are described in Swalen and Gladney[4]; also see Bleaney.[15]

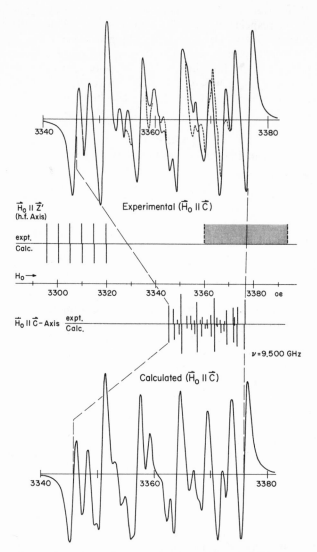

Fig. 6. Comparison of computed and experimental EPR spectra for an ^{27}Al hole center in α-quartz at 77°K and 9.500 GHz. The stick diagrams at center compare the patterns obtained from measured spectra with those generated using MAGNSPEC for $H_0 \parallel z'$ (the quadrupole-hyperfine axis). C-axis spectra synthesized from the experimental and computed stick diagrams using a Lorentzian line of 1 Oe width are shown at the top and bottom, respectively. The broken curve portions of the upper spectrum follow the true experimental spectrum.

with

$$g_\perp = [(g_x \cos \phi)^2 + (g_y \sin \phi)^2]^{1/2}$$

$$g = [(g_\perp \sin \theta)^2 + (g_z \cos \theta)^2]^{1/2}$$

$$A_\perp = \left[\left(\frac{g_x}{g_\perp} A_x \cos \phi \right)^2 + \left(\frac{g_y}{g_\perp} A_y \sin \phi \right)^2 \right]^{1/2}$$

$$K = \left[\left(\frac{g_\perp}{g} A_\perp \sin \theta \right)^2 + \left(\frac{g_z}{g} A_z \cos \theta \right)^2 \right]^{1/2}$$

$$A = \tfrac{1}{4}[(A_\perp A_z/K) - (A_x A_y/A_\perp)]$$

$$B = \tfrac{1}{4}[(A_\perp A_z/K) + (A_x A_y/A_\perp)]$$

$$C = K$$

$$D = (i/4)[A_z g_x g_y g_z(\sin \phi \cos \phi \cos \theta)(A_y^2 - A_x^2)]/K_g A_\perp g_\perp^2$$

$$E = \tfrac{1}{2}[g_\perp g_z(\sin \theta \cos \theta)(A_\perp^2 - A_z^2)]/Kg^2$$

$$F = (i/2)[g_x g_y(\sin \phi \cos \phi \sin \theta)(A_y^2 - A_x^2)]/Kgg_\perp$$

The primes and double primes on the spin components indicate that S and I are quantized in different directions, neither of which is, in general, exactly parallel to the applied field. The quantities g_x, g_y, g_z and A_x, A_y, A_z are the principal (diagonal) components of the tensors.

Computed and experimental[16] spectra of vanadyl etioporphyrin II in castor oil are compared in Fig. 7. The spectrum was fit by an *axial* Hamiltonian for which $g_x = g_y = g_\perp, g_z = g_\parallel, A_x = A_y = A_\perp$, and $A_z = A_\parallel$, with $S = \tfrac{1}{2}, I = \tfrac{7}{2}$. In such a case Eq. (36) can be simplified[17] and the line positions and intensities become a function solely of θ, the angle which the applied field makes to the $\parallel$-axis of the complex. The anisotropic terms are not motionally averaged for this massive complex in the viscous medium, so the powder pattern calculation was used. Only the main ($\Delta m_I = 0$) lines were computed at $10°$ intervals with MAGNSPEC, and the positions and intensities at intermediate orientations were obtained by three-point interpolation. The final powder spectrum is equivalent to 144 orientations (every $\tfrac{5}{8}°$ of arc).

The upper spectrum in Fig. 7 is reproduced from the work of Roberts *et al.*[16] Initially, a fit was attempted using the parameters they specified, $g_\parallel = 1.9474$, $g_\perp = 1.9885$, $A_\parallel = 1.580 \times 10^{-2}$ cm^{-1}, and $A_\perp = 5.40 \times 10^{-3}$ cm^{-1}. Since their spectrometer frequency was not given, it was assigned the tentative value 9.663 GHz on the basis of the fit of the mean g value $(g_\parallel + 2g_\perp)$ to the field scale given. After two adjustments of the parameters the lower spectrum in Fig. 7, which fits the experimental spectrum well, was obtained. The final parameters were $g_\parallel = 1.9590$, $g_\perp = 1.9823$, $A_\parallel = 1.570 \times 10^{-2}$ cm^{-1}, $A_\perp = 5.67_3 \times 10^{-3}$ cm^{-1}. A Lorentzian line (full width = 10 Oe) was used for each component.

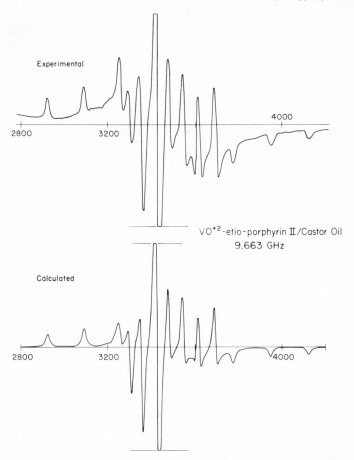

Fig. 7. Comparison of the experimental spectrum of VO^{2+} etioporphyrin II in castor oil (upper curve) with a computed envelope (lower curve) for 9.663 GHz.

More recent work[18] has indicated that the isotropic g value (g_0) in these compounds is probably about 1.9780–1.9790, rather than the value of 1.9748 indicated by the parameters of Roberts *et al.* On the basis we would choose $v_{mr} = 9.6815$ GHz, and obtain about the same fit as in Fig. 7 with $g_{\parallel} = 1.9629$ and $g_{\perp} = 1.9862$.

Case c. A Nitrogen-Related Hole Center in Alkali Silicate Glass*

After x-irradiation, alkali silicate glasses melted under reducing and anhydrous conditions, but in the presence of nitrogen, show an EPR spectrum which can be separated into three component spectra. All three can be

*These results will be presented in more detail elsewhere.

assigned to hole trapping at different centers, point defects or (nonmetal) impurities. Two of these centers can be made inactive by annealing. The third, persisting, center can be related to replacement of an oxygen by a nitrogen in the structure.

A radical structure of the type

$$\overset{\cdot\cdot}{\underset{}{N}}\cdot$$

suggests itself.† Single-crystal ESR spectra of radicals of the same type have been observed‡ [$:\overset{\cdot\cdot}{N}H_2$, $:\overset{\cdot\cdot}{N}F_2$, and $:\overset{\cdot\cdot}{N}(SO_3)_2$]. There an angle α of 100–$110°$ can be deduced, and the appropriate spin Hamiltonian has the form (35) with the parameters

$$g_x \approx g_y \approx 2.003, \qquad g_z \approx 2.006\text{–}2.009$$

$$A_x \approx A_z \approx 0, \text{ and } A_y \approx 30\text{–}50 \text{ Oe}.$$

The experimental spectra of the present nitrogen center in the glass, observed at three irradiation frequencies,** are shown as solid curves in Fig. 8. By systematically adjusting the above approximate parameters (fitting the 9.5-GHz data, then the 16.3-GHz data, with MAGNSPEC + SPREAD), we arrived at the principal values $g_x = 2.0039$, $g_y = 2.0026$, $g_z = (2.0060\ldots$ $2.0135\ldots 2.0240)$ distributed, $A_x \approx A_z \approx 2$ Oe, and $A_y = (36)$ Oe. The particular shape in the bracketed range in the figure could not be approximated with any single value of g_z; only a broad and skewed g_z distribution produced the computed spectra of Fig. 8.

Thus for values of g_z in the range 2.006–2.024 the scalar $\mathscr{H}_{\text{stat}}$ form (36) was used to calculate the line positions for 27 (or 93) uniformly-placed orientations in the first octant (which is the largest region that yields distinct spectra). A three-point interpolation procedure (program SPREAD) was then used to increase this to 345 (or 1529) orientations and a powder spectrum was cumulated with a Gaussian component of 2.0 Oe width. Several spectra were then superimposed with weights indicated in the inset in Fig. 8. These same weights were used at each resonance frequency. The 34.9 GHz, which was obtained after the fitting was complete, shows the long tail required by the broad g_z distribution; the oscillation in the computed curve is a result of not choosing a sufficiently fine distribution. This curve also shows clearly the resolution of g_x and g_y.

---

† Symmetry-related orbitals for such a radical (C_{2v} symmetry) can best be described in a set of reference axes with z along the Si—Si direction, y perpendicular to the plane of the radical, and x along the bisector of the angle α.

‡ For NF_2 see Kasai and Whipple.[19] For $N(SO_3)_2$ see Horsfield et al.[20]

** The spectra at 16.3 GHz were run by E. F. Reichenbecher in the laboratory of Prof. R. Bernheim at the Pennsylvania State University, State College, Pennsylvania. The 35-GHz spectra were run by W. Landraff at Varian Associates, Palo Alto, California.

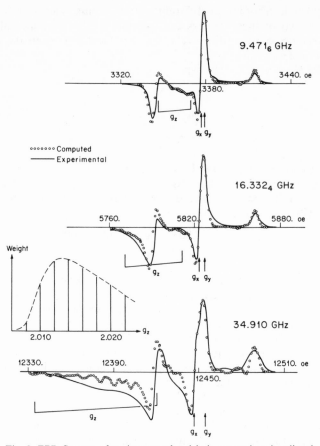

Fig. 8. EPR Spectra of a nitrogen-related hole center in x-irradiated sodium silicate glass for frequencies 9.4716, 16.3324, and 34.910 GHz. The solid curves are measured, the dotted curves are computed. The field positions of the central line of the ^{14}N hyperfine triplet for H_0 along x, y, and z are marked accordingly. The bracketed region denotes the range of g_z values used. The amplitudes used for each g_z are indicated in the inset.

The 34.9-GHz data were obtained primarily for testing the accuracy of the g values derived from the fits at 9.5 and 16.3 GHz. The difference $g_x - g_y$, which had been empirically deduced, is now clearly marked, and the low-field tail is required by the broad g_z distribution. (The oscillations in the computed curve result from the rather coarsely-chosen distribution interval.)

Multispin Systems

Apart from technical storage limitations, systems of several spins can be treated as well. As described in Eq. (34), the present Hamiltonian form has been applied to two hyperfine interactions, one of which, with a third spin I', was isotropic and could be considered as the scalar term

$$JS \cdot S \equiv aS \cdot I$$

If, e.g., several hyperfine tensors are required, the HAMILTON and TENSOR routines are readily expanded to deal with those new tensors A1, A2, etc.

COMMENTS

MAGNSPEC (supplemented by SPREAD) may be regarded as a program for testing spin Hamiltonians against experimental spectra. It has the virtues of flexibility and of using, where possible, straightforward quantum-mechanical methods. Running time, particularly that used by JACOBI, is one important factor in use of the program. No accurate estimate can be made of the execution time for a general problem, but the diagonalization time for HSP is directly proportional to the number of pivotations (elementary rotations) required. This will be reflected in the mixing of the eigenvectors, or the number of large off-diagonal elements in the original matrix, so:

1. The time increases rapidly with matrix dimension.

2. The time roughly doubles for each order of magnitude increase in RATIO, so time can be saved if a less exact diagonalization is requested; this mainly affects the eigenvectors. In addition, only useable field precision (PFIELD) should be requested.

3. The number of lines tracked may be limited by judicious selection of RALLD and JKALLD.

4. In some cases HSP can be made real merely by properly choosing a phase factor; in such cases elimination of imaginary components reduced the number of pivotations by half.

The program* has been assembled and executed on a Control Data 1604A computer with 38K core. Until now no substantial effort has been devoted to using core space efficiently; the program overflows if matrices larger than 24 × 24 complex (or 48 × 48 expanded) are involved. With a little effort we could probably load 30 × 30 or 36 × 36 matrices, but beyond that a larger core would be needed.

This represents a limitation on the type of problem which can be handled if execution time is not excessive. Many EPR problems of interest involve a relatively large number of particles, which are often in symmetry-related groups or have relatively simple interactions. In principle, such problems can be reduced in size below the dimensions of the *direct product* representation used here. Such a reduction is involved in the Wigner–Clebsch–Gordan theory of coupling of two angular momenta L and S in free atoms (see, e.g., Edmonds[21]); in this case of spherical symmetry a transformation from the $|S, S_z, L, L_z\rangle$ representation to a $|S_z, L_z, J, J_z\rangle$ basis, where $\mathbf{J} = \mathbf{L} + \mathbf{S}$, reduces the maximum dimension from $(2S + 1)(2L + 1)$ to $2(L + S) + 1$.

*The source program is basically FORTRAN-II, with little use being made of the additional flexibility allowed by the FORTRAN-63 compiler of the 1604A.

Such reductions are not restricted to the full spherical group or even to point-symmetry operations; the general rule is that any operator $\mathcal{O}$ which commutes with $\mathcal{H}_{stat}$ (i.e., is a constant of the motion) may be used to divide the basis states into nonmixing groups which have different eigenvalues under $\mathcal{O}$. There are only zero matrix elements of $\mathcal{H}_{stat}$ between groups. Such a transformation, illustrated by the atomic vector-coupling coefficients above, is used to secure the proper zero-order linear combinations as a subbasis for each nonmixing group. It may be possible to adapt such methods to reduction of matrix size for computation purposes.

The Hamiltonians [e.g., (17)] used thus far are examples of the Abragam–Pryce[5] two-particle spin system, based on perturbation approximation to the true Hamiltonian. The form of the electronic terms is correct to second order (in L), while the approximation to the nuclear hfs and Zeeman terms is first order. We have found that this Hamiltonian form is sometimes inadequate to fit experimental data for large hfs interactions; there are systematic discrepancies in both the machine and hand perturbation calculations. To correct these discrepancies, use may be made of more accurate forms of the spin Hamiltonian developed recently.[22] In these methods the most general form consistent with the spin multiplicities and symmetry is developed; as a result, the number of parameters is increased and their physical meaning is rendered obscure.

We may also consider possible generalizations of the time-dependent part of $\mathcal{H}_{sp}$. These could be incorporated into our calculations in three ways:

1. In cases of rapid motional narrowing the anisotropic terms in $\mathcal{H}_{stat}$ are averaged (normally to zero). In such cases an isotropic (orientation-averaged) form of $\mathcal{H}_{stat}$ may be used.

2. Such effects as linewidth and shape variations due to relaxation processes could be entered into the SPREAD procedure directly (and rather artificially), or some other statistical averaging[3] may be used.

3. We might consider an extension along the lines of recent theories of relaxation and motional effects[4] by generalizing the time-dependent part of $\mathcal{H}_{sp}$. Suppose the full Hamiltonian for the system is divisible into a static part $\mathcal{H}_0$, a time-dependent part $\mathcal{H}_1$, which is a small perturbation on $\mathcal{H}_0$, and a third (modulation) part which commutes with $\mathcal{H}_0$ but not $\mathcal{H}_1$. Then it is possible to regard $\mathcal{H}_1$ and $\mathcal{H}_{rad}$ on the same basis, normally regarding the motion of $\mathcal{H}_1$ as known only statistically.

ACKNOWLEDGMENTS

The authors appreciate the advice of Professor Aksel A. Bothner-By which led to the spin-manifold representation. They also acknowledge the assistance of Mr. Stephen Ondrey and of Dr. Ronald A. Rutledge and the other personnel of the Mellon Institute Computer Laboratory.

REFERENCES

1. M. Kopp and J. H. Mackey, *J. Computational Physics*, **3**, 539–543 (1969).
2. Leonard I. Schiff, *Quantum Mechanics* (McGraw-Hill Book Co., New York, 1955), Chapters VI and VIII.
3. P. L. Corio, *Structure of High Resolution NMR Spectra*, (Academic Press, New York, 1966).
4. J. D. Swalen and H. M. Gladney, *IBM J. Res. Dev.* **8**, 515 (1964).
5. A. Abragam and M. H. L. Pryce, *Proc. Roy. Soc. (London)* **A205**, 135, (1951); M. H. L. Pryce, *Proc. Phys. Soc.* **63**, 25 (1950).
6. J. H. Wilkinson, *The Algebraic Eigenvalue Problem* (Oxford University Press, Clarendon, 1965).
7. J. Greenstadt, in: *Mathematical Methods for Digital Computers*, A. Ralston and H. S. Wilf, eds. (John Wiley and Sons, New York, 1965), Chapter 7.
8. A. A. Bothner-By and T. Castellano, *J. Chem. Phys.* **41**, 3863–69 (1964).
9. E. O. Schulz-du Bois, *Bell Systems Tech. J.* **38**, 271–290 (1959).
10. D. G. Davis, Mellon Institute, unpublished data.
11. J. H. E. Griffiths, J. Owen, and I. M. Ward, *Defects in Crystalline Solids*, Reports of the Bristol Conference (The Physical Society, London, 1955), p. 81.
12. M. C. M. O'Brien and M. H. L. Pryce, *Defects in Crystalline Solids*, Reports of the Bristol Conference (The Physical Society, London, 1955), p.88.
13. M. C. M. O'Brien, *Proc. Roy. Soc. (London)*, **A231**, 404 (1955).
14. G. S. Smith and L. E. Alexander, *Acta Cryst.* **16**, 462 (1963).
15. B. Bleaney, in: *Hyperfine Interactions*, A. J. Freeman and R. B. Frankel, eds. (Academic Press, New York, 1967), Appendix I.
16. E. M. Roberts, W. S. Koski, and W. S. Caughey, *J. Chem. Phys.* **34**, 591 (1961).
17. B. Bleaney, *Phil. Mag.* **42**, 441 (1951).
18. L. J. Boucher, E. C. Tynan, and Teh Fu Yen, Paper No. 44 at the Pittsburgh Conference on Analytical Chemistry and Applied Spectroscopy, March 4, 1968.
19. P. H. Kasai and E. B. Whipple, *Mol. Phys.* **9**, 497 (1965).
20. A. Horsfield, J. R. Morton, J. R. Rowlands, and D. H. Whiffen, *Mol. Phys.* **5**, 241 (1962).
21. A. R. Edmonds, *Angular Momentum in Quantum Mechanics*, (Princeton University Press, 1957), Chapter 3.
22. G. F. Koster and H. Statz, *Phys. Rev.* **113**, 445 (1959); *Phys. Rev.* **115**, 1568 (1959); T. Ray, *Proc. Roy. Soc. (London)* **A277**, 76 (1964); W. Hauser in: *Paramagnetic Resonance*, Vol. 1, W. Low, ed., (Academic Press, New York, 1963), p. 297.
23. M. S. Itzkowitz, *J. Chem. Phys.* **46**, 3048 (1967).
24. A. G. Redfield, *IBM J. Res. Dev.* **1**, 19 (1957); F. Bloch, *Phys. Rev.*, **105**, 1206 (1957); D. Kivelson, *J. Chem. Phys.*, **33**, 1094 (1960); J. H. Freed and G. K. Fraenkel, *J. Chem. Phys.*, **39**, 326 (1963).

Some Effects of the Sixth Ligand on the Electron Spin Resonance Spectra of Pentacoordinated Low-Spin Co(II) Complexes

Hideo Kon and Nobuko Kataoka

Laboratory of Physical Biology
National Institute of Arthritis and Metabolic Diseases
Bethesda, Maryland

Assignment of the ESR spectra was made for a low-spin pentacoordinated Co(II) complex, Co^{II}(phenylisocyanide)$_5$(ClO$_4$)$_2$, and its hydrated form, in which the sixth coordination site is occupied by a water molecule. It was shown that the occurrence in the frozen state of each form, anhydrous or hydrated, is determined by the solvent, i.e., by whether a hydrophobic or hydrophilic solvent is used. Based upon the assignment, the effect of attaching the sixth ligand on the ESR parameters was examined. It was pointed out that the effect is to shift one of the g components toward a higher value and the ^{59}Co isotropic contact term toward a more negative value. This conclusion was corroborated by examining the ESR parameters of the tetrakis and hexakis (phenylisocyanide) complexes. A brief explanation was given in terms of vibronic mixing of $d_{x^2-y^2}$ into d_{z^2}.

Recently Alexander and Gray[1] showed by ESR and optical absorption that low-spin pentacyanocobaltate(II) ion, $Co(CN)_5^{3-}$, has a square pyramidal, rather than a trigonal bipyramid, structure. The unpaired electron is assigned in $3d_{3z^2-r^2}$ orbital because of the observed g factors, $g_{\parallel} = 1.992 \pm 0.005$ and $g_{\perp} = 2.157 \pm 0.005$. Also, the ^{59}Co hyperfine tensor components were found to be $A = 87 \pm 2$ and $B = 28 \pm 2$ G, respectively, for the directions parallel and perpendicular to the symmetry axis.

An interesting question arises as to whether or not the sixth coordination site of the complex ion is occupied by a molecule of water, especially since the observations were made in aqueous ethylene glycol.[1] In general, when a transition-metal complex has $(3d_{3z^2-r^2})^1$ configuration the magnetic properties are expected to depend upon the presence, absence, or the nature of the sixth (or the axial) ligand in a sensitive way because of the shape of the orbital protruding along that direction. Exactly what changes one can expect in the ESR spectrum, however, by having H_2O at the sixth site has been more or less a matter of speculation.

The present study has been undertaken to investigate the effect of the coordinated H_2O (or the sixth ligand in general) on the ESR parameters of a similar low-spin pentacoordinated system,[2] Co^{II}(phenylisocyanide)$_5$(ClO$_4$)$_2$, in which both anhydrous and hydrated forms were identified by Pratt and

Silverman.[3] These authors observed an ESR absorption at $g = 2.14$ in undiluted powder of the yellow form and one at $g = 2.12$ in the blue form. From these, together with the infrared absorption data, they concluded that the blue powder is in fact the six-coordinated [Co^{II} (phenylisocyanide)$_5H_2O$] complex ion, while the yellow form is its anhydrous modification. The above g values are not very far from the free spin value (2.0023), indicating that both complex ions have an orbitally nondegenerate ground state, and, probably, a square pyramidal structure may be assumed for the anhydrous form.

In this chapter a definite assignment of well-resolved frozen ESR spectra will be made to the yellow anhydrous and blue hydrated ions. Each spectrum does indeed show two g values, which confirms the axial symmetry assumed previously. Furthermore, it is shown that the effect of a water molecule in the sixth site is to shift $g_{\parallel}$ toward higher value and a_{iso} (isotropic hyperfine splitting) due to ^{59}Co to more negative value. This is explained by increased mixing of $3d_{x^2-y^2}$ character into the predominant $3d_{3z^2-r^2}$ by vibronic interaction as the symmetry of the molecule approaches octahedral.

EXPERIMENTAL

Co^{II}(phenylisocyanide)$_5$(ClO$_4$)$_2$ was prepared according to the method of Sacco[2] from phenylisocyanide and Co(ClO$_4$)$_2 \cdot 6H_2O$. The blue powder (hydrated form) thus obtained changes color reversibly to yellow (anhydrous form) in vacuum over P$_2$O$_5$. The elemental analysis of the yellow powder is in good agreement with the calculated values: Co(C$_7$H$_5$N)$_5$(ClO$_4$)$_2$, found: C, 54.94; H, 3.45; N, 8.81; Cl, 9.14; calculated: C, 54.20; H, 3.26; N, 9.06; Cl, 9.17. Melting point, 197–9°. Phenylisocyanide (and C$_7$H$_5$ ^{15}N) was prepared from aniline (aniline-^{15}N) and CHCl$_3$. The CHCl$_3$ used as solvent was dried after washing with H$_2$O over CaCl$_2$ and was distilled over P$_2$O$_5$ several times in diffuse light. CH$_2$Cl$_2$ was dried over CaCl$_2$ and fractionally distilled. Quinoline was dried with fused KOH for a week and similarly distilled. The final water content of the hydrophobic solvents was checked by gas chromatography. All the ESR spectra were recorded as the first derivative using a Varian V-4500 spectrometer with a 9-in. magnet regulated by a Fieldial. Most of the solutions were in the concentration range of 5×10^{-3} to 10^{-2} M. The solution of the yellow anhydrous form in dry quinoline was prepared in a closed system which was attached to a vacuum line. The number of spins was estimated from the integrated intensity of absorption with a Mn^{2+} solution of known concentration as the standard. Degassing the system did not affect the ESR spectra.

ESR SPECTRA

Powder

In polycrystalline powder at room temperature the anhydrous and hydrated forms gave distinct patterns as shown in Figs. 1a and 1b, each showing apparent axial and rhombic symmetry, respectively. At 77°K each

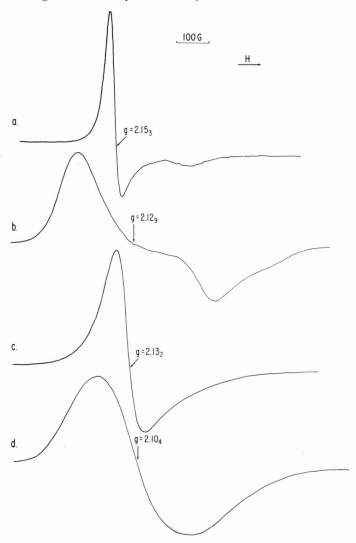

Fig. 1. ESR spectra of $Co(C_7H_5N)_5(ClO_4)_2$ in undiluted powder. (a) Anhydrous, (b) hydrated, at room temperature, (c) Anhydrous, (d) hydrated, at 77°K.

spectrum shows one asymmetrical pattern, but with narrower linewidth in the anhydrous form (Figs. 1c and 1d). The changes in ESR pattern brought by evacuation and replenishing moist air and that due to temperature change are completely reversible, at least for several cycles. The observed apparent g values are indicated in Fig. 1. They are in reasonable agreement with those reported by Pratt and Silverman[3] (spectra not shown) considering the error involved in such broad absorptions.

Frozen Solution

Two different ESR spectra are obtained in frozen solutions (77°K), depending upon the solvent. In a 1:1 mixture of $CHCl_3$ and CH_2Cl_2 the yellow form gives a green solution and on freezing shows the ESR pattern as in Fig. 2a. At this temperature the color of the solution is greenish yellow. It is a well-resolved, typical spectrum for a system with axial symmetry. Interestingly, this spectrum can be obtained regardless of whether the solvents are very well dried or not. Furthermore, the same pattern is obtained from the blue hydrated form when dissolved in the same mixture of solvents.

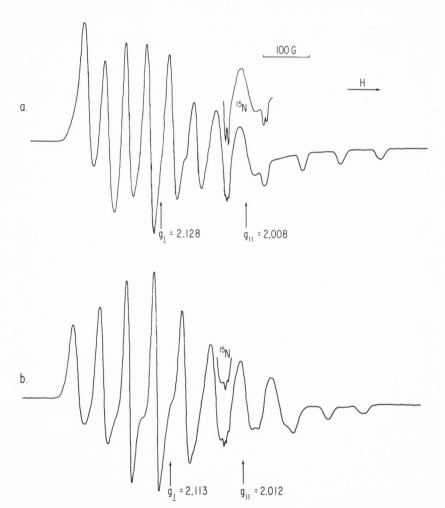

Fig. 2. ESR spectra of $Co(C_7H_5N)_5(ClO_4)_2$ in frozen solutions at 77°K. Solvent: (a) o-nitrotoluene + toluene (1:2), (b) acetone + cyclohexanone (1:1).

Other hydrophobic solvents such as $CH_2Br_2 + CHBr_3(1:1)$, o-nitrotoluene + toluene $(1:2)$ give similar results. Proper selection of mixed solvent is critically important in obtaining a well-resolved frozen pattern.

In spite of the variety of conditions under which the spectrum of the type in Fig. 2a can be obtained, the spectral shape is virtually identical, and shows no indication of other superimposed components. Consequently, it may be concluded that the paramagnetic species in frozen hydrophobic solvents is of only one kind. The fact that the color of the solution changes toward yellow on freezing strongly suggests that this species is the anhydrous form. This is indeed confirmed by the observations, described below, in quinoline and acetic anhydride. The slight blue color still present in frozen solution may appear to represent a small amount of hydrated form, but there are reasons to suspect that it may come from some diamagnetic, e.g., dimerized, species; thus (1) the spectrum does not show any indication of a superimposed spectrum of the blue hydrated form (described below), (2) the spectrum in hydrophobic solvents is found always to account, by the integrated intensity, for only about 80% of the molecules in the system, and (3) when excess phenylisocyanide is added to the same system to convert it to Co^{II}(phenylisocyanide)$_6^{2+}$ ion[4] the integrated intensity increases by roughly 25%, so as to account for all of the complex molecules dissolved. The formation of such diamagnetic species by association is a fairly common observation in low-spin Co(II) complexes.[1,5,6]

In a few hydrophilic solvents such as acetone, methyl ethyl ketone, dimethyl ketone, cyclohexanone, and quinoline the blue as well as yellow powder gives a blue solution, which is unstable at room temperature, as indicated by fading color. When frozen the color is deep blue, and an ESR absorption distinct from that in hydrophobic solvents is obtained, as shown in Fig. 2b. Again the use of a proper mixture of acetone and aryl ketone is needed to obtain good resolution. From the following observations this spectrum can be attributed to the blue hydrated form: The frozen solution of the yellow form in "dry" quinoline gives a spectrum as shown in Fig. 3a, which is interpreted as the spectrum of the anhydrous form (Fig. 2a) plus that in Fig. 2b. On adding to this system a small amount of wet quinoline a pure spectrum identical to the one in Fig. 2b is observed. A similar mixed spectrum is also obtained from cold acetic anhydride solution of blue powder immediately after the solution is made. This spectrum is converted by standing for 30 min at 0° to a spectrum of pure anhydrous form such as in Fig. 2a. This is evidently due to the gradual decomposition of water by acetic anhydride.

All attempts to observe the pure spectrum of the hydrated form in hydrophobic solvents have been unsuccessful because when a small amount of water is suspended in, e.g., $CHCl_3 + CH_2Cl_2$ solution some of the complex molecules are disproportionated to form a hexacoordinated Co^{II}(phenylisocyanide)$_6$ ion[4] instead of the hydrated form. The mechanism by which the hexa(phenylisocyanide) complex is stabilized more than the hydrated complex is conceivably related to the specific way of aggregation of water

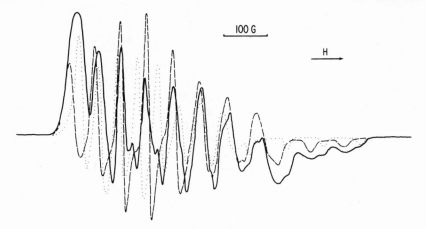

Fig. 3. ESR spectra of $Co(C_7H_5N)_5(ClO_4)_2$ in quinoline at 77°K to show the existence of both the anhydrous and the hydrated forms. Solid curve: in quinoline; dotted curve: in $CHCl_3 + CH_2Cl_2$ (1:1); dashed curve: in acetone + cyclohexanone (1:1). The last two curves are shown to indicate the peak positions; their intensities are not normalized.

molecules in such solvent, but there is no immediate explanation. Also, it has been impossible to observe the pure spectrum of the anhydrous form in hydrophilic solvents due to the difficulties in drying these solvents.

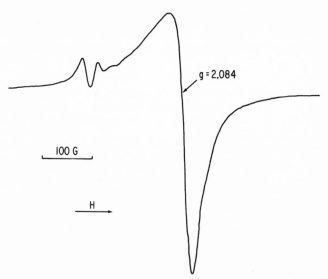

Fig. 4. ESR spectrum of $Co(C_7H_5N)_5(ClO_4)_2$ in dry $CHCl_3$ at room temperature. In hydrophilic solvents practically the same spectra are obtained, except that the small peaks on the low-field side are absent.

Spectra of Liquid Solutions

Since the complexes are unstable in hydrophilic solvents at room temperature, the observations were made below 0°, where the solutions are reasonably stable. In hydrophobic solvents, as shown in Fig. 4, there is one broad, asymmetrical absorption with no hyperfine structure. Also, a small signal with some hyperfine structure (probably due to ^{59}Co) is always present in the low-field side of the main signal in hydrophobic, but not in hydrophilic, solvents. This signal represents only a few per cent of the intensity of the main absorption and may come from some unidentified decomposition product.

The main absorptions, as in some other low-spin Co(II) complexes, do not show isotropic hyperfine splitting of ^{59}Co. The reason for this is discussed in a later section in detail. The asymmetrical line shape may be explained by the variation in linewidth of unresolved hyperfine (^{59}Co) components. However, it must be mentioned that there is no evidence that the spectra in liquid solution represent only one paramagnetic species; from the observations made in the frozen state one may assume that the anhydrous as well as the hydrated form can exist along with some other species, e.g., diamagnetic dimer, in hydrophobic solvents.

The ESR parameters thus obtained are tabulated in Table I together with those of other Co(II) phenylisocyanide complexes.

DISCUSSION

Assignment

From the foregoing observations there is little doubt about the assignment of the two ESR spectra observed in frozen solutions: the absorption from the greenish yellow solution in hydrophobic solvents corresponding to

TABLE I
The g factors and ^{59}Co Hyperfine Splittings (in G) of the Low-Spin Co(II) Phenylisocyanide Complexes

Complex	g_{powder}	$g_{solution}$	g_3	g_2	g_1	$a_{solution}$	A	B
$Co(C_7H_5N)_4X_2$ [a]								
X = Cl	2.116	2.114	2.004	2.156	2.230	(>0)[b]	91.4	$0 > B > -3$[c]
Br	2.089	2.082	2.008	2.090	2.175	(>0)[b]	96.7	$0 > B > -5$[c]
I	2.058	2.036	2.005	(2.052)[d]		—	96.7	—
$Co(C_7H_5N)_5(ClO_4)_2$								
Anhydrous	2.153	2.083	2.008	2.126	2.126	(-5)[b]	81.2	-47.8
Hydrated	2.129	2.080	2.012	2.113	2.113	(-16)[b]	71.3	-59.4
$Co(C_7H_5N)_6(ClO_4)_2$ [e]	—	2.065	2.014	2.085	2.085	-28	64.0	-75.0

[a]N. Kataoka and H. Kon, to be published.
[b]Estimated from A and B.
[c]Estimated from the linewidths.
[d]Calculated from $g_{solution}$ and g_3, assuming axial symmetry.
[e]N. Kataoka and H. Kon, J. Am. Chem. Soc. **90**, 2978 (1968).

the anhydrous form, while the one from deep blue solution in hydrophilic solvents is that of the hydrated form. Since the spectrum of the anhydrous form is obtained even from the blue powder dissolved in hydrophobic solvents, it must be assumed that a dehydration process has taken place during freezing. A possible explanation of this seemingly unusual process would be that in hydrophobic solvents the freezing out of H_2O molecules becomes the dominant process over the hydration of the complex as the temperature is lowered, thereby shifting the equilibrium toward dehydration of the complex.

It may be pointed out that the sixth coordination site of the anhydrous form is not occupied by the solvent molecule, since scarcely any change in spectral pattern has been observed in various solvents. The coordination of perchlorate ion may probably also be ruled out, because Pratt and Silverman[3] concluded from infrared absorption data that ClO_4^- is not coordinated in powdered form even in the absence of H_2O molecule.

Both spectra (Figs. 2a, b) show the typical aspect for a randomly-oriented system with axial symmetry; the parallel and perpendicular bands appear to be split into eight hyperfine components by a ^{59}Co nucleus ($I = \frac{7}{2}$), although some of them are obscured by overlapping. The superhyperfine coupling (4.8 G in hydrophobic and 5.3 G in hydrophilic solvents) with one ^{14}N nucleus (confirmed by ^{15}N labeling) is resolved in one component ($m = +\frac{1}{2}$) of the parallel band. This fact, combined with the observed $g_{||}$ values (2.008 and 2.012 for anhydrous and hydrated forms, respectively) being sufficiently close to the free spin value (2.0023), indicates that the unpaired electron orbital is predominantly of d_{z^2} symmetry. Since in a trigonal bipyramid the unpaired electron is expected to be in degenerate orbitals, it is concluded that the complex in anhydrous form has indeed a square pyramidal structure, in contrast to Co^I (methylisocyanide)$_5ClO_4$, which is known to be a trigonal bipyramid.[6] The hydrated form, then, must be a distorted octahedron. Apparently, the orientations of the phenyl groups of isocyanide do not, in this case, disturb the axial symmetry of the molecule.

Effect of the Sixth Ligand

There are two ESR parameters, $g_{||}$ and a_{iso} (isotropic metal hyperfine splitting), which appear to reflect the effect of the sixth ligand (H_2O molecule in the present case) in the axial position. A component of the g factor is expressed in general by

$$g_{ij} = 2.0023 + \sum_n [\xi \langle 0|L_i|n \rangle \langle n|L_j|0 \rangle]/(E_0 - E_n)]$$

in the perturbation treatment, which is a valid approximation for the present case. Here ξ is the spin-orbit coupling constant in the metal complex, $\langle n|$ the excited state coupled, through the i component of the angular momentum operator L_i, to the ground state in which the unpaired electron is in the d_{z^2} orbital, and E_n and E_0 are the respective energies.

In the present approximation the d_{z^2} orbital is not coupled to any other orbital in the d shell when $i = j = z$, and therefore $g_\parallel = 2.0023$. The positive deviation observed in the hydrated form, however, is small but significant. Similar positive shifts in $g_\parallel$ have been observed for low-spin Co^{II} (phthalocyanine) with the axial positions occupied by various N-containing bases,[8] and also in the hexacoordinated Co^{II} (phenylisocyanide)$_6$(ClO_4)$_2$.[4]

An explanation analogous to the one given for the hexacoordinated molecule by the present authors[4] should also apply to this case, i.e., when the symmetry of the molecule approaches the octahedral by attaching the sixth ligand there can be a small amount of mixing of $d_{x^2-y^2}$ into the predominantly d_{z^2} orbital as the result of vibronic interactions.[9] The ground state would then be

$$|0\rangle = \cos^2(\phi/2)(d_{z^2}) + \sin^2(\phi/2)(d_{x^2-y^2})$$

where ϕ is a function of time. Since the $d_{x^2-y^2}$ orbital can be coupled by L_z to the lower d_{xy} orbital and cause a positive shift of the order of 0.3 in $g_\parallel$ as observed in many Cu(II) complexes, the small positive deviations in the systems being discussed can be easily accounted for by a small fraction of $d_{x^2-y^2}$ character in the unpaired electron orbital. The same mixing should also affect $g_\perp$, but this vibronic effect would be masked by that due to the change (by adding the sixth ligand) in the energy levels of d_{xz}, d_{yz} orbitals to which d_{z^2} is coupled through $L_{x,y}$.

Another effect of the sixth ligand may be found in the isotropic metal hyperfine splitting a_{iso}, although this is not actually resolved in the present complexes. In fact, the latter is a peculiar situation observed in several other low-spin Co(II) complexes. Thus in $Co(CN)_5^{3-}$ ion in aqueous solution there is only one absorption, with peak-to-peak width of about 80 G; also, in Co^{II} (phenylisocyanide) $_4X_2$, where X is Cl, Br, or I, there are large halogen isotropic splittings in liquid solution, but no Co hyperfine structure. It has been shown for these tetrakis(phenylisocyanide) complexes that the isotropic contact term has positive sign.[10]

The isotropic metal contact interaction in a transition-metal complex is in general the result of spin imbalance (χ) in the metal s orbitals caused by the unpaired spin in the d shell, i.e., the spin polarization effect. It is known that in the majority of cases studied so far χ has a negative sign and is surprisingly insensitive to the change of electronic state of the metal ion or the variation of the ligands, its magnitude being approximately 3 in atomic units.[11] For ^{59}Co, $\chi = -3$ corresponds to about 90 G splitting. The exceptional positive sign in the cases cited above is best interpreted as the result of the small s character in the unpaired electron orbital. Thus when the symmetry of the complex ion allows, d_{z^2} and s orbitals can form a linear combination, and because of the large probability density of s orbitals at the nucleus, a small amount of s character would cause a considerable positive contribution to the spin imbalance. A simple estimation would show that 18% of 4s plus 3% of 3s mixed to d_{z^2} orbital gives a positive contribution of about 100 G to ^{59}Co splittings and will override the negative component due to spin polarization.[10] The small a_{iso} values resulting from near cancellation of the two contributions would be one of the reasons for the absence of isotropic ^{59}Co hyperfine structure in various low-spin Co(II) complexes in solution. In Co^{II}(phenylisocyanide)$_5$(ClO$_4$)$_2$ and its hydrated form similar fractional s character could also be present. However, if we estimate the contact term from the equations for the parallel (A) and perpendicular (B) hyperfine couplings [e.g., Equation (9) of McGarvey[11]], the isotropic contact terms for the anhydrous and the hydrated forms turn out to be negative (-21.3 and -30.5 G, respectively), indicating that the fractional s character decreases from tetrakis through pentakis to the hydrated pentakis(phenylisocyanide) complex. This trend is consistent with the order of the positive shift in $g_\parallel$ discussed above and suggests that the fractional s character decreases to the extent that the $d_{x^2-y^2}$ character is introduced by vibronic interaction as the complex ion tends to be more octahedral. In fact, the one nearest the octahedral in this series of complexes, Co^{II}(phenylisocyanide)$_6$(ClO$_4$)$_2$, shows the largest $g_\parallel(2.014)$ and the most negative contact term (-39.8 G). By the same reasoning the positive contact term and smaller deviations of $g_\parallel$ in Co^{II}(phenylisocyanide)$_4$X$_2$ (where X is halogen) indicate the more distortion from octahedral symmetry in the halides, probably because of the mostly ionic nature of Co$-$X bonding. If we can extend the same reasoning to Co(CN)$_5^{3-}$, in which $g_\parallel$ is very close to 2.0023* and the contact term is a small negative quantity (-11.4 G), it may be concluded that either a water molecule is not coordinated at the sixth site, or the bonding is perhaps much weaker than Co$-$CN bondings, so that the effective symmetry of the molecule is still far from octahedral.

Recently, Pratt and Silverman[12] concluded, mainly from the similarity of the optical absorption spectrum of pentacyanocobaltate ion and that of the pentaisocyanide complexes in the hydrated form, that in aqueous solution the pentacyanocobaltate ion has the hexacoordinated structure, [Co(CN)$_5$H$_2$O]$^{3-}$. It must be emphasized that our conclusion is based entirely upon the ESR observations in frozen solution, and in no way contradicts the loose hydration or the rapid exchange of H$_2$O between the bound and the free states in liquid solution.

REFERENCES

1. J. J. Alexander and H. B. Gray, *J. Am. Chem. Soc.* **89**, 3356 (1967).
2. A. Sacco, *Gazz. Chim. Ital.* **84**, 370 (1954).
3. J. M. Pratt and P. R. Silverman, *Chem. Comm.* **1967**, 117; *J. Chem. Soc. (A)* **1967**, 1286.
4. N. Kataoka and H. Kon, *J. Am. Chem. Soc.* **90**, 2978 (1968).
5. H. Kon, *Spectry. Letters* **1**, 49 (1968).
6. L. Malatesta and A. Sacco, *Gazz. Chim. Ital.* **83**, 499 (1953).
7. A. Cotton, T. G. Dunne, and J. S. Wood, *Inorg. Chem.* **4**, 318 (1965).
8. J. M. Assour, *J. Am. Chem. Soc.* **87**, 4701 (1965).
9. M. C. M. O'Brien, *Proc. Roy. Soc. (London)* **A281**, 323 (1964).

*Our own measurement in aqueous solution gave $g_\parallel = 2.003 \pm 0.001$, $g_\perp = 2.176 \pm 0.001$, which are somewhat different from those by Alexander and Gray.[1]

10. N. Kataoka and H. Kon, to be published in *J. Phys. Chem.*
11. B. R. McGarvey, *J. Phys. Chem.* **71**, 51 (1967).
12. J. M. Pratt and P. R. Silverman, *J. Chem. Soc.* (*A*), p. 1291 (1967).

Metalloproteins as Studied by Electron Paramagnetic Resonance*

J. Peisach†
Departments of Pharmacology and Molecular Biology
Albert Einstein College of Medicine
Yeshiva University, New York, New York

and

W. E. Blumberg
Bell Telephone Laboratories
Murray Hill, New Jersey

Low-temperature EPR spectroscopy has been used to study copper- and iron-containing proteins. The information derived from these studies includes the oxido-reduction function of the metal, special features of the ligand field of the metal especially as they pertain to the protein environment, and interactions between paramagnetic centers. Correlations which exist between spectrophotometric and magnetic properties are related to the spin states of paramagnetic species. These are used to describe spin exchange phenomena in such systems as oxyhemoglobin, oxyhemocyanin, cytochrome c oxidase, copper uroporphyrin, and various model metal complexes.

INTRODUCTION

EPR has been successfully used in the study of the biochemistry of metalloproteins both as a means of elucidating certain physical and chemical properties of metal sites and also as an indicator of the participation of various paramagnetic metal ions in the function of these proteins.‡ Applications of EPR to biochemistry can be divided into two classes. The first includes those studies that give qualitative information concerning the presence of paramagnetic metal ions and the changes that they may undergo, without consideration of quantitative information concerning the chemical physics relating to their physical environments. The second, and these are the ones to which we will address ourselves, include those studies which *do* yield information concerning the specific physical environments of metal ions in metalloproteins. EPR signals have been detected for vanadium, titanium,

*This investigation was supported in part by a Public Health Service Research Grant (GM-10959) from the Division of General Medical Sciences. This is Communication No. 116 from the Joan and Lester Avnet Institute of Molecular Biology.

†Career development awardee of the United States Public Health Service (1-K3-GM-31,156) from the National Institute of General Medical Sciences.

‡For a recent review see Beinert and Palmer.[1] Some of the more recent work in the field is discussed briefly in the book edited by Ehrenberg et al.[2]

manganese, cobalt, iron, copper, and molybdenum in materials of biological origin. Iron, copper, and molybdenum are most often found in paramagnetic proteins, and therefore these have been the elements studied most intensively. Only in the cases of iron and copper have experiments been performed sufficiently physical in nature that quantum-mechanical descriptions of their results could lead to physical information concerning the environment of metal sites. We will give several examples of these studies. All of these examples have one feature in common which simplifies the analysis of the experimental data: no crystal field splitting is present in the EPR spectra. This means that all spectra are interpretable in terms of an effective spin of $\frac{1}{2}$. This being true, the magnetic field H where resonant absorption occurs is directly proportional to the microwave frequency v according to the equation $hv = g\beta H$, where h and β are Planck's constant and the Bohr magneton, respectively. Carried over from Zeeman spectroscopy in the gaseous phase, g is sometimes called the spectroscopic splitting factor, and is a numerical constant characteristic of the electronic distribution giving rise to the EPR signal. In the solid state g is formally a second-rank tensor, which means that it can have three different values, g_x, g_y and g_z, along three mutually perpendicular directions in the solid. If as is often the case, the x and y directions are physically indistinguishable, the symmetry is said to be axial, and g can have only two independent values, $g_\perp$ in the xy plane and $g_\parallel$ in the z direction. When the electron distribution containing the unpaired spin residues partially on an atom the nucleus of which has a magnetic moment, the interaction between the spin and the nuclear moment, called the hyperfine interaction, gives rise to a splitting of the EPR absorption. The multiplicity into which the absorption is split is equal to $2I + 1$, where I is the value of the nuclear spin in units of Planck's constant. The spacing of this splitting (A) depends on the size of the nuclear moment and details of the distribution of electronic spin within the atom. The following section gives a very elementary summary of the relevant aspects of EPR spectroscopy of iron and copper atoms. More complete discussions may be found elsewhere.[3,4]

THE STATES OF COPPER AND IRON FOUND IN BIOLOGICAL SYSTEMS THAT GIVE RISE TO EPR SIGNALS

The only state of copper which is found in samples of biological origin and gives an EPR absorption is Cu(II). The electronic configuration of this species is d^9 and it has a single electronic spin. The stereochemistry of most Cu(II) complexes of high stability requires them to be square planar or tetragonally-distorted octahedral with four close-lying ligands and two which are further removed. The Jahn–Teller effect does not permit an octahedral environment for the d^9 configuration to remain cubic. These arrangements place the unpaired spin in the $d_{x^2-y^2}$ antibonding orbital, which has an axially symmetrical EPR spectrum, with $g_\parallel$ greater than $g_\perp$. In addition to the g values, the EPR spectrum of Cu(II) is characterized by hyperfine structure arising from the interaction with Cu(II) nuclei. Typically, for the $d_{x^2-y^2}$ orbital $A_\parallel$ gives well-resolved structure and $A_\perp$ does not. A typical EPR

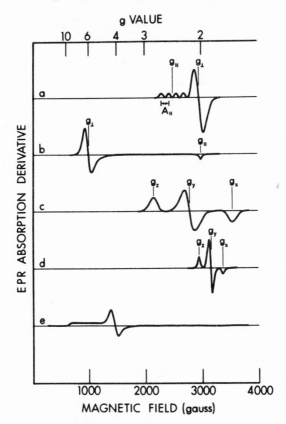

Fig. 1. Hypothetical EPR spectra. (a) Cu(II) (d^9, $S = \frac{1}{2}$);
(b) high-spin ferric heme (heme d^5, $S = \frac{5}{2}$); (c) low-spin
ferric heme (Heme d^5, $S = \frac{1}{2}$); (d) non-heme iron as found
in ferredoxin; (e) non-heme Fe(III) as in ferrichrome,
transferrin, and rubredoxin, where Fe(III) is an octahedral
environment having a rhombic distortion.

derivative spectrum for Cu(II) is shown in Fig. 1a. In the case of such a well-
resolved spectrum the magnetic parameters of the spin-Hamiltonian equation
for Cu(II) can be obtained directly from the spectrum. The $g_\perp$ is obtained from
the ratio of the microwave frequency ν to the value of magnetic field H at the
zero crossing in the spectrum, $g_\parallel$ is obtained from the ratio of the microwave
frequency to the magnetic field at the center of the four-line parallel hyperfine pattern, $A_\parallel$, in energy units, is the spacing between components of the
hyperfine pattern, measured in gauss, multiplied by $g_\parallel \beta$, and $A_\perp$ is unresolved.

Iron exists in several different states observable by EPR. Two of the most
common of these arise from Fe(III) in porphyrin structures as are found in
heme proteins. The electronic configuration of Fe(III) in hemes is d^5 either
in the high-spin $e_g^2 t_{2g}^3$ ($S = \frac{5}{2}$) or the low-spin t_{2g}^5 ($S = \frac{1}{2}$) forms. In both these

forms four of the ligands for the metal are nitrogen atoms from porphyrin (x, y directions), and the remaining z ligands are variable and establish the spin state of iron. In the high-spin case (Fig. 1b) the z ligands contribute little to the ligand field of the iron, and the symmetry of the system is axial or nearly axial. The axial crystal field splitting is always so large (10–30 cm^{-1}) that EPR transitions are only observed between members of the lowest-lying doublet. These are characterized by $g_{\parallel} = 2$ and $g_{\perp} = 6$.

In the low-spin case (Fig. 1c) the contributions of the z ligands to the ligand field of the iron are large, and the symmetry of the system is nearly cubic. The Jahn–Teller effect prevents exact cubic symmetry, and low-spin Fe(III) EPR signals are often resolved into three components g_x, g_y, and g_z, with g values ranging from approximately 1 to 3. Only one of these is allowed to be less than 2.

Yet another form of iron having an EPR spectrum is found in the class of non-heme iron-containing proteins (Fig. 1d). This type of iron signal was first observed in a biological system and has never been exactly reproduced with model compounds. Thus the electronic structure of the state giving rise to it remains unknown. The EPR of this non-heme iron has characteristic g values close to that of the free electron, and does not arise from one of the common d-electron configurations of a single iron atom. A discussion of possible structures giving rise to this resonance will be given below.

Another form of Fe(III) which has an EPR spectrum is found in that class of compounds where Fe(III) is in an octahedral environment having a rhombic distortion (Fig. 1e). A typical EPR signal has a strong absorption near $g = 4.3$ and a weak absorption extending from approximately $g = 1$ to $g = 10$, which is most easily observed at low temperatures. Here the crystal field splitting varies from 2 to 5 cm^{-1} and is significantly weaker than is observed for heme iron. Typical proteins where this signal is observed include transferrin, ferrichrome, and rubredoxin. The reader will find a more complete discussion of this resonance elsewhere.[5]

Since the most-prevalent isotopes of iron have no nuclear magnetic moment and occur in total natural abundance of 97.8%, Fe(III) EPR signals lack hyperfine structure. Attempts to produce hyperfine splitting by introducing ^{57}Fe are made difficult by the very small moment of this species. Yet information can be obtained from such experiments in certain cases, and these will be discussed below.

TURACIN, COPPER UROPORPHYRIN III

Although not a metalloprotein, copper uroporphyrin III or turacin, a pigment naturally occurring in the feathers of the South African touraco, has been studied with EPR[6] in order to answer various questions about this material: (1) What is the physical state of turacin as it occurs in the bird's feather? (2) What is the state of aggregation of this material in a dispersion and in a solution in the presence of various amines which appear to affect its solubility and optical properties? Also, turacin was considered sufficiently interesting for study as a model compound for possible electronic configurations for Cu(II) in copper-containing proteins.

Turacin is soluble at high pH, but is insoluble in acid medium. The optical spectrum of acidified turacin can be determined using gelatin as a dispersant. In acid, turacin exhibits optical maxima at 383, 540, and 584 mμ. The addition of NH_3-water to precipitated turacin dissolves it, yielding a solution with optical maxima at 398, 526, and 562.5 mμ. Any organic amine, except for methylamine, shifts the optical spectrum of this material to longer wavelengths. A large effect is observed with caffeine, where the optical spectrum of turacin exhibits maxima at 405, 528.5, and 564 mμ.

EPR studies of turacin under these different conditions yield three entirely different spectra. A solution of turacin to which ethylamine has been added exhibits the EPR spectrum shown in Fig. 2. This same spectrum is observed with caffeine and a host of other organic amines. This spectrum is typical for Cu(II) in a planar environment and exhibits superhyperfine structure attributable to the electronic interaction of a single Cu(II) with four equivalent ^{14}N nuclei. In contrast to the hypothetical example above (Fig. 1a), only two of the four peaks of the $A_\parallel$ pattern are resolved, and analysis of such an overlapping spectrum is more difficult than in the hypothetical case. This EPR spectrum for turacin exhibits no interactions between paramagnetic species and represents a species uniformly distributed in solution, i.e., Cu(II) uroporphyrin III monomers.

The EPR of turacin in ammoniacal solution is given in Fig. 3. This EPR spectrum is not interpretable in terms of a spin Hamiltonian for a single Cu(II), but is interpretable as representing two Cu(II) atoms interacting at a short distance.

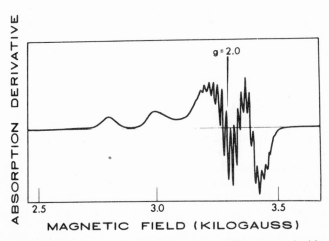

Fig. 2. EPR spectrum of turacin in aqueous medium saturated with ethylamine. Spectrum of 10 mM of turacin taken at 77°K. (After Blumberg and Peisach.[6])

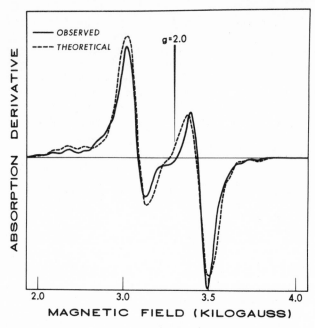

Fig. 3. EPR spectrum of turacin in NH_3-water. The solid curve shows the observed EPR absorption derivative of 10 mM turacin in 2 M NH_3 at 77°K. The dashed curve was computed with spin-Hamiltonian parameters derived from the spectrum. (After Blumberg and Peisach. [6])

    The spectrum shows the effect of the dipole–dipole interaction between the two Cu(II) spins and also the sharing of the hyperfine interaction of the Cu(II) atoms equally between the two spins. The simplest form of such an interaction occurs when the porphyrin molecules are arranged parallel one to another, with the Cu atoms arranged on a line perpendicular to both planes. The best theoretical EPR spectrum for two interacting Cu(II) atoms in this arrangement is obtained when the paramagnetic centers are approximately 3.5 Å apart. Thus under the conditions of these solutions turacin exists only as a dimer and not as any other oligomeric form.

    In the presence of acid, turacin precipitates, and the EPR spectrum of this material (Fig. 4) is entirely different from those observed for the monomer or the dimer. Here there is a narrow, almost symmetrical absorption which occurs when there is an exchange interaction shared among many Cu(II) atoms. Magnetic dipolar interactions among the Cu(II) spins (dashed line, Fig. 4) would not lead to the EPR spectrum observed. Since the line shape is narrow and symmetrical in the precipitated material, the interactions among the Cu(II) atoms must take place among molecules of many different orientations in order to average all components of the g tensor. Thus the crystalline structure is not strictly speaking an infinite linear array. Since this spectrum and also the one for the dimer does not

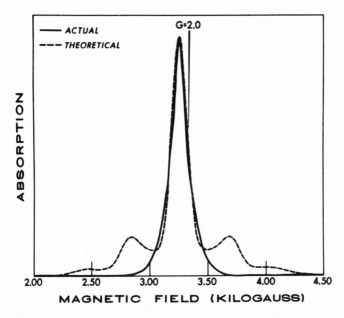

Fig. 4. EPR spectrum of turacin precipitated by acid. The solid curve shows the EPR absorption spectrum of 5 mM turacin precipitated by 2 M acetic acid. The dashed curve is the computed spectrum of a linear array of Cu(II) with no exchange coupling included. (After Blumberg and Peisach.[6])

lose its intensity as the temperature is lowered to 1.5°K, the isotropic part of the interaction between adjacent molecules is not greater than approximately 1 cm^{-1}. Since the spectrum observed for acidified turacin is consistent with what one would expect from an approximately linear array of weakly exchange-coupled Cu(II) atoms, it is ascribed to a polymeric or aggregate form of turacin molecules which are arranged one on top of another in microcrystalline form. Thus one concludes that when acid is added to a solution of turacin there is a large aggregation of molecules and the formation of turacin polymer. The conversion of monomer to dimer and ultimately to polymer is directly associated with a downward shift of Soret peaks and an upward shift of visible peaks of the optical spectrum, and there is a direct relation between optical spectra and the state of aggregation as determined by EPR.

It was observed that a 10^3 M excess of caffeine was effective in preventing the precipitation of turacin in acid medium. Here it is asserted that caffeine prevents this precipitation by forming weak chemical bonds with turacin by means of π-orbital overlap between caffeine and porphyrin. This physically prevents the interaction of turacin molecules which would be necessary for polymer formation. Turacin can be precipitated at neutral pH by high concentration of KCl. This suggests that the dipolar interaction of water with the hydrophylic ionized carboxyl groups of the molecule are necessary for solubility. Once the water is effectively removed, either by turacin protonation or by dipolar binding to salts, the turacin molecules interact with each other and precipitate from solution. The precipitation mechanism probably involves aggregation of dimers, since acid precipitation is retarded by those materials which aid in monomer formation.

Finally, the red feathers of the touraco bird were examined with EPR. The spectrum obtained is precisely the same as observed for the polymer, and therefore the turacin in the feather is in its polymerized form. This suggests that during the laying down of the copper porphyrin in the feather barbs and barbules turacin is precipitated either by acidification or by high salt concentration. This means then that rather than being dispersed throughout the feather keratin, in the manner of a dye, turacin is aggregated in specific locations.

STELLACYANIN

A number of the proteins which contain copper are intensely blue, i.e., molar extinction coefficients in the visible red region can be over 4000 per copper.[7] This is much higher than is observed for copper chelates. Also, the EPR spectra of these blue proteins is distinctive and has not been reproduced in model compounds. One of these proteins, ceruloplasmin, contains both Cu(I) and Cu(II), and it was suggested that the intensity of color is due to charge-transfer transitions between them.[8] Subsequently, a copper protein was isolated from *Pseudomonas* which contained a single Cu(II) per molecule and yet was intensely blue.[9]

It was suggested on theoretical grounds that the unusual optical and magnetic properties of the blue copper proteins is due to various distortions away from the typical square planar environments usually found in copper chelates. The study described here was performed with stellacyanin,[10,11] a mucoprotein isolated from *Rhus vernicifera*, containing a single Cu(II) per molecule, in which various molecular constraints on the copper site of the molecule could be relaxed stepwise and thus is consistent with the theoretical description of the unusual copper site.

The optical spectrum of stellacyanin is given in Fig. 5. This spectrum is typical for the blue copper protein ceruloplasmin,[5] laccase,[10] ascorbate oxidase,[11] and plastocyanin.[12] Here, however, and in the case of plastocyanin, all three peaks in the visible are extremely well resolved, probably reflecting a greater rigidity in the molecule with regard to random strain in the ligand field of the copper. The optical spectrum of stellacyanin remains constant in the pH range 2.0–8.3, and also in the presence of such protein denaturants as urea, guanidine, and mercuric chloride. ORD studies of stellacyanin indicate that at the position of each peak in the visible there is a Cotton effect indicative of a noncentrosymmetrical environment associated with the chromophore.

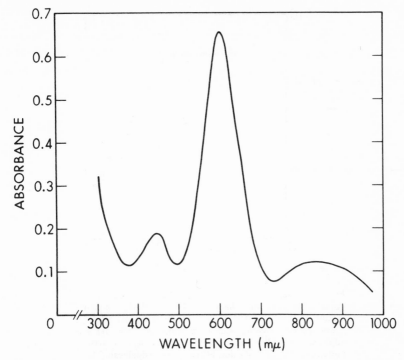

Fig. 5. Optical absorption spectrum of stellacyanin in the visible and near-ultraviolet. The protein concentration was 0.165 mM; the solvent was 5 mM Tris-acetate buffer, pH 5.4. (After Peisach et al.[11])

As the pH is raised above 8.3 the color of stellacyanin changes from blue to purple and eventually to light pink. This change is indicated in Fig. 6. As for the major visible peak, similar shifts in optical spectrum with pH are observed with the absorptions at 448 and 845 mμ. Despite the shift in wavelength maxima, the intensity of the peaks in the pH range 2.0–11.5 remains virtually unchanged.

Returning the purple or pink stellacyanin solutions to below pH 8.3 restores the optical spectrum of the molecule so that it again appears blue. The reversibility of the color-change phenomenon declines with time, so that if stellacyanin is maintained at pH 11.5 for 20 min, only 85% of the original color is restored. Above pH 11.5 the intensity of color diminishes quickly, and lowering pH after decolorization restores neither the original spectrum nor the intensity of color.

Figure 7 shows the EPR spectrum of stellacyanin taken at various hydrogen-ion concentrations. The sample was 0.5 ml of stellacyanin (1.35 mM) containing 85 μg of copper per ml in 0.05 M Tris-acetate buffer, pH 5.4. The sample was frozen in liquid nitrogen and was examined at pumped liquid-helium temperature (approximately 1.5°K). After EPR observations

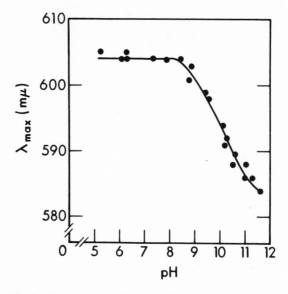

Fig. 6. pH dependence of the position of the major visible optical absorption of stellacyanin. Three milliliters of stellacyanin, 0.175 mM, in 0.05 M Tris-acetate buffer, pH 5.4, were contained in a cuvette with 1-cm light path. The pH was adjusted by adding concentrated NaOH with a syringe microburet, so that dilution effects were insignificant. After pH adjustment the position of the major visible peak was measured with a Cary model 14 spectrophotometer. (After Peisach et al.[11])

were made the pH was adjusted to 10.8 with 6 M NaOH, and the sample was quickly refrozen. The same procedure was followed for pH 11.5. For the pH 12.0 sample NaOH was added at 0° until the solution was an intense pink, and the sample was immediately frozen. After the EPR observation was made the sample was thawed, upon which rapid decolorization occurred. The pH was then found to be 12.0 as measured at room temperature. Upon refreezing of this solution the EPR spectrum was found to be unchanged and was identical to that of *bis*-biuret copper complex in basic solution. The spectrum shown here at pH 5.4 does not change in the pH range 2.0–8.3. The spectrum represents the EPR absorption of a single Cu(II) with $g_x = 2.03$, $g_y = 2.06$, and $g_z = 2.30$. The most unusual feature of the spectrum is the very rhombic nature of the hyperfine interaction, $A_x = 20 \pm 30$, $A_y = 140 \pm 30$, and $A_z = 100 \pm 30$ (A_x and A_z were unresolved at 9000 MHz). Raising the pH above 8.3 changes the EPR spectrum, until eventually at pH 12 it is indistinguishable from the EPR spectrum of *bis*-biuret-Cu(II) in basic medium. The EPR changes seen in the pH range 8.3–12 are precisely where the changes in optical spectrum described above occur. Upon thawing a stellacyanin solution which had been brought to

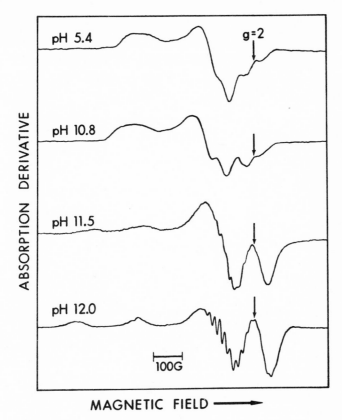

Fig. 7. X-band EPR spectra for stellacyanin at various values of pH.
(After Peisach et al.[11])

pH 12.0 immediately before freezing, a rapid decolorization takes place, but no change in the EPR spectrum is observed when the sample is reexamined at 1.5°K.

Below pH 2.0 the intensity of the visible spectrum diminishes and copper is released by the stellacyanin molecule (Fig. 8). If stellacyanin is incubated at pH 1.1 at 22° for 24 min, approximately 80% of its color disappears, as monitored at 604 mμ. Concomitant with this decrease of optical absorbance is the decrease of protein-bound copper to about 60% of the original. Raising the pH to 7.3 at this point causes an immediate but slight increase in color followed by a slow increase to about 65% of the original. Introduction of EDTA before neutralization prevents recolorization except for the small, fast initial increase. It is noteworthy that EDTA has no affect on the native material. Incubation at pH 1.1 for longer time periods results in a diminished ability to recolorize at pH 7.3 and an increase of release of copper by the protein. The denaturation experiments with stellacyanin at high and low

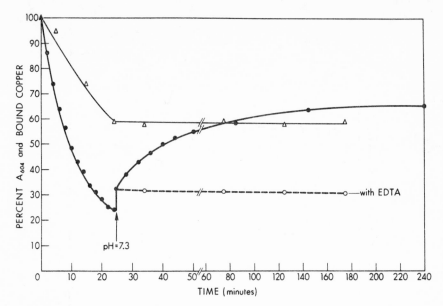

Fig. 8. The effect of acid on the blue color and the release of copper from stellacyanin. Stella-cyanin, 0.1 ml of a 1.0 mM solution, was added to 0.90 ml of HCl, pH 1.0, and the absorbance at 604 mµ (darkened circles) was read periodically. After 24 min (vertical arrow) 0.5 ml of phosphate buffer, pH 11.5, was added to half the sample. The final pH was 7.3. To the other half of the sample 10 µl of 0.1 M EDTA were added before the addition of phosphate. The A_{60} was followed for the sample with EDTA (open circles) and for the sample without EDTA (darkened circles). At the times indicated the percentage of bound copper (triangles) was calculated from the amount of DCO-reacting copper (sample without EDTA). (After Peisach et al. [11])

pH can be combined to describe certain aspects of the chemical physics of the copper site.

The three d–d transitions in the visible spectrum of symmetrical copper complexes are electronically forbidden. In stellacyanin, however, these transitions are partially allowed (up to 3%) by certain odd components in the ligand field of the copper. These odd components can arise only if there are constraints on the ligand atoms of the copper arising from the structure of the protein. A release of these constraints by a change in configuration of the protein would cause the odd components to vanish and the transitions to become highly forbidden. Therefore the intensity of blue color is one in-dication of the integrity of the molecule. The fact that the denaturation of the copper site in the protein proceeds via several steps when the pH is either increased (Figs. 6, 7) or decreased (Fig. 8) from the range of stability allows us to make a very general model of the binding site described below.

Previous calculations based on ligand field theory[15] have shown that approximately all the binding sites for divalent copper can be represented as modifications of an arrangement of four strong ligands in a square and two weaker ligands on an axis perpendicular to the square. This ideal symmetrical arrangement, when subjected to several distortions represented mathematically by ligand field potential functions, can be made to reproduce the physical properties of almost all copper complexes. In the case of stellacyanin it is not possible to specify a unique set of these distortions, but one can say that on the basis of the EPR, optical absorptions, and ORD spectra there must be at least one rhombic distortion and at least one odd distortion. The rhombic part must be complex with respect to x and y axes passing through the strong ligand atoms of the undistorted square. The odd distortion must include at least some $Y_3^2 + Y_3^{-2}$ or "saddle field" and some Y_1^0 or $Y_1^1 + Y_1^{-1}$, which displace the copper atom from the center of the square. Structure I in Fig. 9 shows a schematic arrangement of the four strong ligand atoms. The central sphere represents the divalent copper atom, the surrounding four spheres the ligand atoms to which the copper is attached. Shown are the hypothetical displacements R to make the rhombic field, and further displacements O to make the odd field, which includes the saddle field and the out-of-center components. These displacements must be made at the expense of d–electron ligand-field energy and therefore must be maintained by constraints in the structure of the protein. When these constraints are destroyed by denaturation of the protein structure these displacements will vanish. When the pH is increased above 8.3 but not above 11.5 the EPR absorption (Fig. 7) changes from a rhombic to an axial spectrum, showing that the constraint giving rise to the R displacements of structure I (Fig. 9) is relaxed. However, since the optical absorptions (Fig. 6) are still intense, the O displacements have not vanished. This is the situation prevailing in structure II of Fig. 9. Above pH 11.5 the optical absorption intensity decreases to about the same as that for the bis-biuret-Cu(II) complex, showing that the O displacements have nearly vanished. This relaxation of the odd components would not affect the EPR spectrum.

Structure III, which represents stellacyanin at high pH, is the square planar complex of copper and peptide nitrogen found for most proteins at high pH. This fact is confirmed by the nitrogen superhyperfine structure of the EPR of structure III. Since the changes between structures I and II and between II and III are interpreted as a relaxation of constraints on the ligand field and not by an interchange of ligand atoms, the four strong ligand atoms in native stellacyanin must be four nitrogens.

The low-pH denaturation of the copper binding site in stellacyanin proceeds via an entirely different pathway. Figure 8 shows that during acid denaturation there are four types of copper present. The first step in the acid denaturation is the release of the O constraint as the optical absorptions of the copper decrease toward an unmeasurable intensity, and is represented by structure IV of Fig. 9. Otherwise, the ligand field remains much as before, since the copper is still not vulnerable to chelating agents such as EDTA. The renaturation of structure IV is represented by the small increase in $A_{604\,m\mu}$ immediately upon neutralization of an acidified stellacyanin solution (arrow, Fig. 8) in the absence or presence of EDTA. It is not known whether the R displacement should remain in structure IV or not. EPR studies would

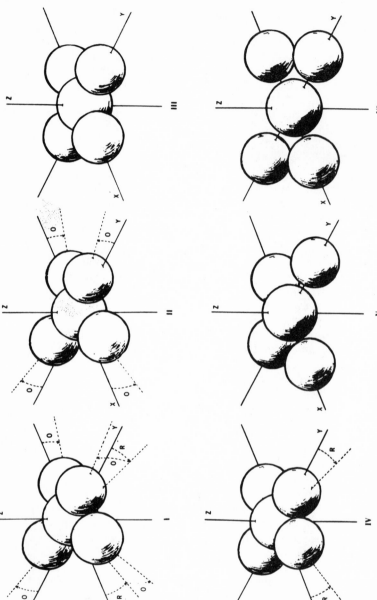

Fig. 9. Diagrammatic sketch of the four strong ligand atoms of the Cu binding site of stellacyanin. Structure I is the native site. Structure II is the intermediate stage in high-pH denaturation (pH 8.3–11.5), where only the R displacement has been relaxed. Structure III is the final stage in high-pH denaturation above 11.5, where both R and O have been relaxed. Structure IV is the first stage of acid denaturation, pH below 2.0, in which the O displacement has been relaxed, but the Cu is not accessible to chelating agents. Structure V is the second stage in acid denaturation, in which the Cu, although renaturable, is accessible to chelating agents. Structure VI is the last stage in acid denaturation, in which the Cu is not renaturable and is accessible to DCO. (After Peisach et al [11])

resolve this question, but would be difficult to carry out, as the concentration of structure IV is never very large during the acid denaturation process.

Further denaturation of the native ligand field of stellacyanin leads to structure V, in which two ligand positions are so loosely bound that EDTA can compete with them for the copper, but not so loosely bound that chelating agents can remove the metal ion. The EDTA does not remove the copper from this configuration, but does prevent renaturation when the pH is raised to the stable region. In the absence of EDTA, when the pH is raised structure V renatures slowly, with a half-time of about 20 min.

The final stage in acid denaturation is represented by structure VI, in which all four strong ligand atoms are so loosely bound that the copper can spontaneously dissociate from the protein or can react with chelating reagents such as dicyclohexanoneoxalyldihydrazone (DCO). When the denaturation has reached structure VI no renaturation can occur.

Thus stellacyanin represents an example of a copper-containing protein in which the copper is situated in an unusual, strained environment. Since similarities occur in optical spectra, ORD, and EPR between stellacyanin and ceruloplasmin, ascorbate oxidase, laccase, and plastocyanin, it is suggested that similar strained ligand-field environment for Cu(II) exists in these molecules.

Oxyhemocyanin has the same characteristic optical properties enumerated above.[16] However, this molecule is diamagnetic. It exhibits no low-temperature EPR and has no paramagnetic susceptibility, even at room temperature. On the basis of optical properties alone one would ascribe a d^9 electronic configuration to copper. It is well known that a single oxygen molecule combines per two copper atoms of hemocyanin. The apparent diamagnetism is due then to the spin coupling of two closely-neighboring Cu(II) atoms directly or *via* a pair of oxygen atoms.

When the two unpaired electrons [in this case those from Cu(II)] appear on closely-neighboring atoms in the same structure there can exist a finite amount of electronic overlap between them. Of the two possible spin states, parallel and antiparallel, the electronic coupling in the latter case is antiferromagnetic and the two spins combine to exactly cancel their paramagnetism. As an example, the spins of two copper atoms in dimeric copper acetate are coupled in this manner, but to a lesser extent. One may write, then, the following reaction for the oxygenation of hemocyanin as

$$2\,Cu(II) + \cdot O\!-\!O\cdot \rightarrow Cu(I)\ {}^-O\!-\!O^-\ Cu(I)$$

Structures in which oxygen acts as an electron acceptor have been postulated and discussed for oxyhemoglobin and oxymyoglobin.[17]

HEMOGLOBIN

Hemoglobin, the oxygen-transporting molecule found in mammalian erythrocytes, is composed of four individual heme-containing subunits. The molecule in its active form contains Fe(II) in the high-spin form (S = 2).

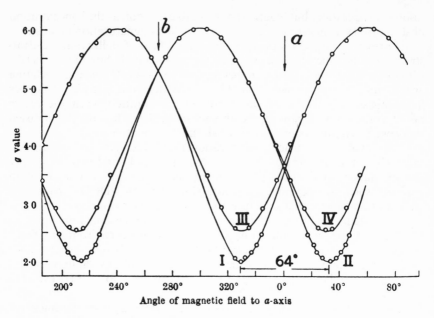

Fig. 10. Variation of g values of the four high-spin heme iron atoms with the direction of the magnetic field in the (001) plane of horse methemoglobin. (After Ingram et al.[18])

Oxidation of hemoglobin yields methemoglobin, in which the iron is now ferric and high-spin ($S = \frac{5}{2}$). It is this species that exhibits an EPR spectrum typical for high-spin heme iron, with effective spin $\frac{1}{2}$ and with g values of 2, 6, and 6 (see Fig. 1b). Each of the hemes in the tetramer is paramagnetic, and their reactive orientations one to another are determined by the conformation of the protein to which they are attached. Methemoglobin can be crystallized, and this material has been used to study the relative orientations of the planes of the four hemes in the tetrameric molecule. The elegance and simplicity of these experiments[18] will be described.

A single crystal of methemoglobin was placed in an EPR spectrometer along the (001) face. The magnet was rotated and the g values were measured as a function of angle of rotation (Fig. 10). Another crystal was then placed in the spectrometer, but now along the (110) face, and the same experiment was repeated. Since g for an electronic state with effective spin of $\frac{1}{2}$ is a tensor, it has the geometrical representation of an ellipsoid having principal axes equal to the three principal g values, and intermediate values are calculable from the solid geometry. If the axis of the heme porphyrin plane are defined by Cartesian coordinates (Fig. 11), then the measured g values at any angle of rotation of the magnet are related to the principal g values by the following equation:

$$g^2 = g_z^2 \cos^2\theta + g_x^2 \sin^2\theta \cos^2\psi + g_y^2 \sin^2\theta \sin^2\psi$$

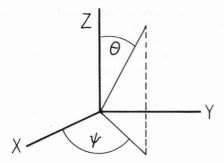

Fig. 11. Coordinates of a single heme porphyrin of hemoglobin. Here θ is the polar angle from the normal to the heme plane and ψ is the azimuthal angle from a fixed point in the heme plane.

These two experiments provided two different cross sections of the g ellipsoids for the four heme groups of hemoglobin, and since the principal axes are known, they provide a complete description of the orientation of these ellipsoids.

The crystal placed along the (001) axis had measured g values varying from a maximum of 6.0 to a minimum of 2.0 for one pair of hemes and from 6.0 to a minimum of 2.55 for another pair of hemes (Fig. 10). All four hemes exhibited maximal g values of 6, since any arbitrary plane must pass through all four heme planes ($\theta = 90°$). The pair of hemes with g minima of 2.0 have g minima (at $\theta = 0°$) separated by 64°, and their z axes, perpendicular to the two hemes, are located in the (001) plane $\pm 32°$ from the crystalline area. Those with measured minima of 2.55 are located so that the normal to the heme plane makes an angle of $\pm 77°$ with the [001] direction. Therefore the normal to the heme plane lies 13° above and below the (001) plane.

Although the results of this study do not define in absolute terms the relation of any or all of the hemes to particular parts of the protein, they do place a restriction on possible orientations the hemes may have one to another. Since there is no spin–spin interaction between the porphyrins of methemoglobin, both in the crystal and in solution, it is concluded that the porphyrins are well separated in the molecule. Furthermore, these results, as for many studies of protein crystals, only define an orientation in the crystal which may not be the same in solution.

The addition of azide to methemoglobin converts the heme to the low-spin form ($S = \frac{1}{2}$). EPR experiments were performed with crystalline methemoglobin azide[19] as before. In this case the principal g values are converted from 2, 6 and 6 to 1.72, 2.22, and 2.80. Here the porphyrin no longer provides the iron with an environment having fourfold symmetry, as g_x no longer equals g_y. The nonequivalence of x and y in the heme plane is brought about by asymmetrical π-bonding to both proximal histidine and to azide. Similar analysis as in the high-spin case indicates that the orientation of the principal axes of the g tensors are the same as before. This signifies that the conversion of high-spin methemoglobin to its low-spin azide derivative does not change the relative orientation of the hemes and the polypeptide chains of the molecule. When hemoglobin is oxygenated the crystalline material differs significantly from crystalline deoxyhemoglobin in its x-ray pattern. These changes cannot be studied by EPR, as the ground states of the hemes are not suitable. Unfortunately, the binding of azide by methemoglobin, which has been studied, does not bear a resemblance to the oxygen-binding mechanism of hemoglobin.

NON-HEME IRON PROTEINS

In many non-heme iron-containing proteins, derived both from plant and from animal sources, an EPR signal was detected by Beinert[20] with two g values near to but less than 2. This signal, which was from the beginning attributed to iron, was observed only under reducing conditions in biological materials containing more than one iron atom per molecule. Ever since its initial discovery biochemists have tried to ascertain whether the signal is indeed due to iron, and if so, what special ligand arrangements are responsible for it, since no ligand field acting on the five d electrons of low-spin iron can give two g values less than 2. It had also been noted that when iron is removed from some of those materials yielding this EPR signal either reversibly or irreversibly, a quantity of H_2S was also lost. This sulfur could only be derived from inorganic sulfide ("labile sulfide") which is an integral part of the metalloprotein.[21] Furthermore, reconstitution experiments could be successfully performed only with the addition of both iron salts and H_2S.[22] The following questions are then raised about these non-heme iron-containing proteins:

1. Is the EPR signal observed actually produced by unpaired electron spins localized at the iron?
2. Is sulfur a necessary ligand for the iron in order to produce such signals?
3. What is the arrangement of atoms necessary to produce such signals?

Attempts to answer these questions have been made using EPR.

However, in order to answer them, only low-molecular-weight molecules in which the single redox-active material is non-heme iron were used. These include the non-heme iron-containing proteins from *Azotobacter vinelandii* containing two iron atoms per molecule,[21] a protein from *Clostridium* (two iron atoms), ferredoxins from *Clostridium pasteurianum* (seven iron atoms),[24] spinach (two iron atoms),[29] and adrenodoxin (two iron atoms)[26] derived from bovine adrenal cortex. The labile sulfide in many non-heme iron proteins is equimolar to the iron contained by them.

In order to determine whether iron is actually responsible for the EPR signal, isotopic labeling experiments were performed where ^{57}Fe was substituted for ^{56}Fe. For the *Azotobacter* protein[23] the microorganism was grown in an isotopically-enriched medium, and in the case of spinach ferredoxin[25] the iron was chemically exchanged. In the former case a broadening of all the components of the EPR spectrum by about 20 G shows that an isotropic interaction with one or more ^{57}Fe nuclei has been introduced. The most obvious conclusion is that the signal arises from an unpaired electron which is indeed principally localized on iron.

In another experiment ^{57}Fe was chemically exchanged with ^{56}Fe of adrenodoxin[27] (Fig. 12). Here Beinert and his associates tried to reproduce the observed EPR spectrum of the isotopically-substituted protein by assuming various values for the hyperfine coupling constants and assuming that either one or two Fe atoms were involved. They found that the spectrum could only be reproduced with two iron atoms having equal hyperfine constants $A = 0.0012$ cm^{-1}, and they interpreted this finding as indicating that the EPR signal of the redox-active site of adrenodoxin contains two iron atoms. As far as the unpaired spin is concerned, these two iron atoms are electronically equivalent, but that does not necessarily mean that both iron atoms are chemically equivalent in the structure. [Electron equivalence has been demonstrated for Cu(II)-glycylglycylhistidylglycine, where the ligands

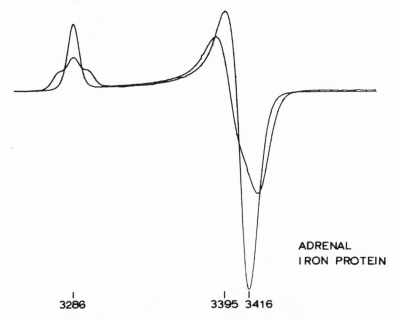

ADRENAL
IRON PROTEIN

|
3286   3395 3416

Fig. 12. EPR spectrum of adrenodoxin. The g values of the native material containing ^{56}Fe are approximately $g_\parallel = 2.02$ and $g_\perp = 1.93$. Chemical substitution of ^{57}Fe splits the $g_\parallel$ into triplet components having ratios of $1:2:1$. The splitting at $g_\perp$ is unresolved.

of Cu(II) are clearly two different types of nitrogen atoms.[28]] The large A value observed in adrenodoxin indicates that the unpaired electron contains a large amount of $4s$ character on each iron atom, as it is larger than the maximum A value which can be produced by a single d electron through core polarization.

Isotopic experiments with spinach ferredoxin[25] also confirm that iron is associated with EPR signal. The experiments did not allow the determination of the number of iron atoms as did the adrenodoxin experiments.

Another type of substitution was performed where ^{33}S was substituted for the naturally-occurring ^{32}S isotope in *Azotobacter* iron protein.[29] Here, too, the microorganism was grown in an isotopically-enriched medium and the final preparation contained 52% isotopic substitution. A line broadening of approximately 9 G in $g_{\parallel}$ and 4 G in $g_{\perp}$ was observed. It is concluded from these experiments that one or more sulfur atom(s) is associated with the iron responsible for the EPR signal, and that sulfur is a necessary ligand for the iron.

In order to determine whether labile sulfur is actually a necessary ligand for iron, ^{77}Se was substituted for labile ^{32}S in reconstituted adrenodoxin. The EPR signal is shown in Fig. 13. Analogous to the ^{57}Fe substitution, this spectrum can be reproduced assuming isotropic interactions with two

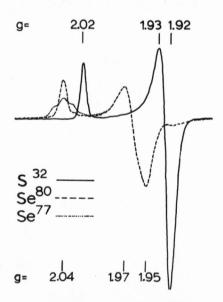

Fig. 13. EPR spectrum of adrenodoxin having selenium substituted for "labile sulfide." With ^{80}Se the $g_{\parallel}$ is shifted to 2.04. With ^{77}Se this part of the spectrum is split into triplet components having ratios of $1:2:1$. The splitting at $g_{\perp}$ is unresolved.

equivalent ^{77}Se nuclei. Here the interaction constant was $A = 0.0010\,\text{cm}^{-1}$. One concludes from these experiments that indeed the EPR signal is derived from an unpaired spin of a molecular orbital comprising at least two iron atoms and two sulfur atoms. These experiments do not distinguish between the sulfur dissociating from the molecule with the removal of iron (labile sulfide) and that which remains in cysteine of the polypeptide chain.

Since these EPR signals obtained with non-heme iron-containing proteins had never been seen before, it is difficult to ascribe structures responsible for them. Various iron model compounds have been studied, but none of them produce the same EPR as observed in the metalloproteins, and none of them contain sulfur as ligands for the iron. Three models have been suggested for the iron in the various non-heme iron-containing proteins. The first[30] is a nonspecific model in which it is not necessary to consider two iron atoms to yield molecular orbitals having the required properties. Here it is stated that the EPR can arise from the perturbation of an unpaired spin in a molecular orbital, not necessarily of sulfur, by a diamagnetic transition-metal ion. This general model does not of necessity require iron, and can be demonstrated by other metal ions. The system consisting of low-spin Co(III) and p-phenylenediamine radical ion yields these resonances, for example.

A second model is a two-iron-atom model suggested by Brintzinger et al.[31] Here it is suggested that the structure giving rise to the EPR signal of non-heme iron protein consists of two low-spin iron atoms in a tetrahedral ligand field, one d^5, having the electronic configuration $(S = \frac{1}{2})$, and the other d^6 $(S = 0)$, each ligated to the same two sulfur atoms. Possible resonance forms for this model include Fe(I). Similar EPR signals could be observed with a bis-hexamethylbenzene-iron complex, where it is assumed that the principal electronic configuration of iron is d^7 $(S = \frac{1}{2})$.[32]

The third model, suggested by Gibson et al.,[33] comprises two iron atoms. Here it is assumed that each has the equivalent configuration of d^5 in the oxidized state, and the two are coupled antiferromagnetically to make them have a diamagnetic ground state. In the reduced state one of the atoms is reduced to d^6 $(S = 2)$ and the other remains d^5 $(S = \frac{5}{2})$. The antiferromagnetic coupling now produces a total spin of $\frac{1}{2}$ for the two iron atoms.

In both the Brintzinger and the Gibson models it is assumed that the unpaired spin giving rise to the EPR is primarily a d electron, while the Blumberg and Peisach model assumes that the unpaired spin giving rise to the EPR is primarily a σ or π electron in the ligand field of the iron atom or atoms. None of these models lead to a definitive description of the redox-active site of non-heme iron-containing proteins. The answer will be determined by knowing the exact structure of the site either from chemical analysis or by physical means.

A first approximation is based on the isotopic substitution experiments indicating that two iron atoms and two sulfur atoms make up the redox-active site in non-heme iron-containing proteins. Although this site can yield the EPR signals described here, it is believed that these spectra may also be observed with other metal–ligand combinations not occurring in non-heme iron-containing proteins. Metal pairs are required in the proteins to produce the physiologically significant redox potentials for these systems.

CYTOCHROME c OXIDASE

Another problem of current interest concerns the paramagnetic centers of cytochrome c oxidase. This metalloprotein is the terminal oxidase of the mammalian electron-transport chain, i.e., it mediates the oxidation of reduced cytochrome c and also the reduction of oxygen to water. It is the former reaction which has been studied in great detail,[34] yet the mechanism of the latter reaction is not as yet understood.

Cytochrome c oxidase is a complex molecule made of two reversibly separable subunits, cytochrome a and cytochrome a_3. The native oxidase, isolated from mammalian mitochondria, contains two hemes (heme a and heme a_3) and two copper atoms.[35] In the native protein each of these metallo components can react with specific reagents; e.g., heme a_3 combines with cyanide or CO, while heme a does not. The anaerobic reduction of cytochrome c with DPNH and a synthetic electron acceptor, phenazine methosulfate, shows that a total of four reducing equivalents are required to reduce cytochrome c oxidase.

EPR studies performed on cytochrome c oxidase indicate that in the completely oxidized material, containing two Cu(II) and two Fe(III), signals attributable to only a single Cu(II) and a single heme Fe(III) $(S = \frac{1}{2})$ are observed.[36] Anaerobic reductive titrations with phenazine methosulfate and DPNH (Fig. 14) show a decrease in both Cu(II) and Fe(III) $(S = \frac{1}{2})$ signals and the appearance of new heme EPR signals attributable to both high-spin Fe(III) $(S = \frac{5}{2})$ and low-spin Fe(III) $(S = \frac{1}{2})$ species. Experiments performed by quickly mixing and freezing cytochrome c oxidase and reduced cyto-chrome c or dithionite show a much larger high-spin heme signal than with phenazine methosulfate titration. The addition of cyanide to the mixture converts all of the new Fe(III) signal to a new low-spin signal, while fluoride converts all of the new signal to the high-spin signal. The total quantity of new heme species observed by EPR under partial reduction of cytochrome c oxidase with any particular concentration of reductant is independent of whether it has been converted to the low- or high-spin form. It is clear, how-ever, that the new high-spin and low-spin heme species observed arise from an iron atom which does not contribute to the EPR in the native material.

As was mentioned before, the EPR signals attributable to copper represent approximately one-half the copper in cytochrome c oxidase. At no time during the anaerobic reduction of the protein does this Cu(II) signal increase, nor is there the appearance of any new Cu(II) species. This can only mean that the initial reduction of native cytochrome c oxidase takes place at the Cu(II) site which does not contribute to the EPR signal. Although this metal species has an odd number of electrons, its EPR spectrum is not observed. A possible explanation for this lack of EPR spectrum may be that

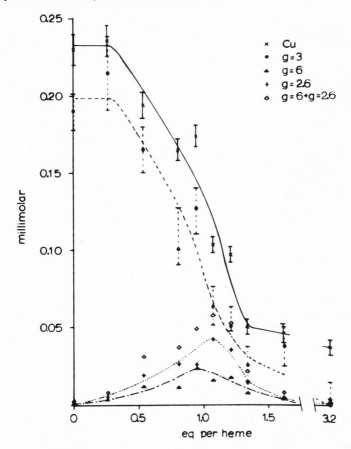

Fig. 14. EPR signal vs equivalents of DPNH added during the anaerobic reduction of cytochrome c oxidase. Spectra were taken near liquid-nitrogen temperature. (After Beinert.[35])

there is an antiferromagnetic coupling of the non-EPR detectable Cu(II) with those paramagnetic heme species that only appear after partial reduction of the molecule. Reduction of this Cu(II) to Cu(I) breaks the coupling and causes the appearance of the new heme EPR signals described above. The fact that both high- and low-spin heme signals appear suggests that both can exist in the native material in an environment in which they are readily interconvertible. The addition of covalent ligands (e.g., CN^-) or ionic ligands (e.g., F^-) firmly establishes one or the other spin states of this heme.

Once the electron enters the cytochrome c oxidase molecule at the non-EPR detectable Cu(II) site it is quickly redistributed to those paramagnetic metal ions which gave the EPR signal in the native material. This is known from the fact that the fast addition of reductant and freezing yielded much larger high-spin Fe(III) heme signals than those observed under steady-state conditions. The reductions in low-spin heme and in Cu(II) signals of the native material with titrant appear to parallel one another, suggesting equal affinity of both species for the reducing equivalent contributed by the Cu(I).

The interaction of numerous inhibitors and chelating agents with the metal atoms of cytochrome c oxidase have been studied. Both heme a_3 and the copper of cytochrome a_3 are freely accessible to these reagents, while heme a and copper of cytochrome a are not. It has been suggested that the EPR detectable Cu(II) signal arises from the copper associated with cytochrome a, and the EPR nondetectable copper is located in cytochrome a_3. If it is the latter component of cytochrome c oxidase which both receives electrons and is also effected by various metal ligands, it is suggested that the cytochrome c oxidase molecule is composed of a and a_3 subunits with the metal ions of cytochrome a_3 located together on the outside of the molecule and the metal ions of cytochrome a separated and buried on the inside of the molecule. If, however, the EPR detectable Cu(II) arises from the copper of cytochrome a_3, then the Cu(II) of cytochrome a is the initial recipient of electrons from reductants, and it is the metal ions of cytochrome a which reside on the outside of the molecule of cytochrome c oxidase. The choice of which model is correct will be based on EPR spectroscopic experiments combined with fast kinetic studies.

CONCLUSIONS

In conclusion, we have indicated the types of EPR studies of metalloproteins which can lead to chemical-physical information concerning the metal sites in these molecules. In no way do we wish to imply that these are the only EPR experiments of this type, nor do we wish to imply that only this type of experiment yields useful information concerning biological materials. Experiments in which EPR signals appear, disappear, or change are often helpful in describing the molecular structure and function of metalloproteins. However, none of these latter studies yield information regarding the chemical physics of the metal ion sites as do the experiments discussed in this chapter.

ACKNOWLEDGMENT

The authors wish to thank Drs. H. Beinert and W. H. Orme-Johnson for their data supplied on selenium-substituted adrenodoxin prior to publication.

REFERENCES

1. H. Beinert and G. Palmer, *Advan. Enzymology* 27, 105 (1965).
2. A. Ehrenberg, B. G. Malmström, and T. Vänngård, eds., *Magnetic Resonance in Biological Systems* (Pergamon Press, Oxford, 1966).

3. G. E. Pake, *Paramagnetic Resonance* (W. A. Benjamin, New York, 1962).
4. W. Low, *Paramagnetic Resonance in Solids* (Academic Press, New York, 1960).
5. W. E. Blumberg, in: *Magnetic Resonance in Biological Systems*, A. Ehrenberg, B. G. Malmström, and T. Vänngård, eds. (Pergamon Press, Oxford, 1966), p. 119.
6. W. E. Blumberg and J. Peisach, *J. Biol. Chem.* **240**, 870 (1965).
7. W. E. Blumberg, W. G. Levine, S. Margolis, and J. Peisach, *Biochem. Biophys. Res. Commun.* **15**, 277 (1964).
8. W. E. Blumberg, J. Eisinger, P. Aisen, A. G. Morell, and I. H. Scheinberg, *J. Biol. Chem.* **238**, 1675 (1963).
9. H. S. Mason, *Biochem. Biophys. Res. Commun.* **10**, 11 (1963).
10. J. Peisach, W. G. Levine, and W. E. Blumberg, in: *Magnetic Resonance in Biological Systems*, A. Ehrenberg, B. G. Malmström, and T. Vänngård, eds., (Pergamon Press, Oxford, 1966), p. 199.
11. J. Peisach, W. G. Levine, and W. E. Blumberg, *J. Biol. Chem.* **242**, 2847 (1967).
12. T. Nakamura and Y. Ogura, in: *The Biochemistry of Copper*, J. Peisach, P. Aisen, and W. E. Blumberg, eds. (Academic Press, New York, 1966), p. 389.
13. C. R. Dawson, in: *The Biochemistry of Copper*, J. Peisach, P. Aisen, and W. E. Blumberg, eds., (Academic Press, New York, 1966), p. 305.
14. W. E. Blumberg and J. Peisach, *Biochim. Biophys. Acta* **126**, 269 (1966).
15. W. E. Blumberg, in: *The Biochemistry of Copper*, J. Peisach, P. Aisen, and W. E. Blumberg, eds., (Academic Press, New York, 1966), p. 40.
16. K. E. Van Holde, *Biochemistry* **6**, 93 (1967).
17. J. Peisach, W. E. Blumberg, B. A. Wittenberg, and J. B. Wittenberg, *J. Biol. Chem.* **243**, 1871 (1968).
18. D. J. E. Ingram, J. F. Gibson, and M. F. Perutz, *Nature* **178**, 906 (1956).
19. J. F. Gibson and D. J. E. Ingram, *Nature* **180**, 29 (1957).
20. H. Beinert, in: *Non-heme Iron Proteins: Role in Energy Conversion*, A. San Pietro, ed. (Antioch Press, Yellow Springs, Ohio, 1965), p. 23.
21. W. Lovenberg, R. B. Buchanan, and J. C. Rabinowitz, *J. Biol. Chem.* **238**, 3899 (1963).
22. R. Malkin and J. C. Rabinowitz, *Biochemistry* **6**, 3880 (1967).
23. Y. T. Shethna, P. W. Wilson, R. E. Hansen, and H. Beinert, *Proc. Natl. Acad. Sci.* **52**, 1263 (1964).
24. G. Palmer, R. H. Sands, and L. E. Mortenson, *Biochem. Biophys. Res. Commun.* **23**, 357 (1966).
25. G. Palmer and R. H. Sands, *J. Biol. Chem.* **241**, 253 (1966).
26. G. Palmer, H. Brintzinger, and R. W. Estabrook, *Biochemistry* **6**, 1658 (1967).
27. H. Beinert, W. H. Orme-Johnson, and R. E. Hansen, *Ann. N.Y. Acad. Sci.*, in press.
28. G. F. Bryce, *J. Phys. Chem.* **70**, 3549 (1966).
29. T. C. Hollocher, F. Solomon, and T. E. Ragland, *J. Biol. Chem.* **241**, 3452 (1966).
30. W. E. Blumberg and J. Peisach, in: *Non-heme Iron Proteins: Role in Energy Conversion*, A. San Pietro, ed. (Antioch Press, Yellow Springs, Ohio, 1965), p. 101.
31. H. Brintzinger, G. Palmer, and R. H. Sands, *Proc. Natl. Acad. Sci.* **55**, 397 (1966).
32. H. Brintzinger, G. Palmer, and R. H. Sands, *J. Amer. Chem. Soc.* **88**, 623 (1966).
33. J. F. Gibson, D. O. Hall, J. H. M. Thornley, and F. R. Whatley, *Proc. Natl. Acad. Sci.* **56**, 987 (1966).
34. E. C. Slater, B. F. Van Gelder, and K. Minnaert, in: *Oxidases and Related Redox Systems*, T. E. King, H. S. Mason, and M. Morrison, eds., (John Wiley and Sons, New York, 1965), p. 667.
35. H. Beinert, in: *The Biochemistry of Copper*, J. Peisach, P. Aisen, and W. E. Blumberg, eds. (Academic Press, New York, 1966), p. 213.
36. B. F. Van Gelder, W. H. Orme-Johnson, R. E. Hansen, and H. Beinert, *Proc. Natl. Acad. Sci.* **58**, 1073 (1967).

The Electron Spin Resonance Spectrum of Copper-Doped Palladium bis-Benzoylacetonate Crystals*

Michael A. Hitchman and R. Linn Belford

Noyes Chemical Laboratory and Materials Research Laboratory
University of Illinois
Urbana, Illinois

The EPR spectrum of about 0.5% Cu^{2+} diluted into the two crystalline forms of Pd bis-benzoylacetonate is reported. The spectrum of the $(11\bar{1})$ face of the needle modification shows an anomalous doubling of the EPR lines which is explained by the presence of twinning in this crystallographic form. The principal g values and directions have been obtained in terms of an arbitrary molecular coordinate system using a specially written computer program. The following parameters result: $g_x = 2.049_0$; $g_y = 2.047_8$; $g_z = 2.242_8$; $A = -181 \times 10^{-4}\,cm^{-1}$, $B = -28 \times 10^{-4}\,cm^{-1}$; with the in-plane g axes lying approximately between, rather than along, the copper–oxygen bonds. These results are used to estimate bonding parameters for the molecule. For comparative purposes bonding parameters for Cu bis-acetylacetonate have been recalculated from the EPR data of Maki and McGarvey using recent estimates of the d-level energies. While the σ-bonding molecular orbital coefficients are almost the same in the two chelates, both the in-plane and out-of-plane π-bonding coefficients are significantly smaller in the complex with the phenyl-substituted ligand. The orientation of the g tensor implies that the along-the-bonds rhombic effect of the phenyl groups does not affect the orientation of the (xz) and (yz) orbitals, and causes no significant contamination of the ground-state orbital by the $(3z^2 - r^2)$ orbital, which can therefore be ignored in deriving the covalency parameters.

INTRODUCTION

Paramagnetic resonance measurements can yield valuable information on the effective ligand field symmetries and orbital geometry as well as on bonding in transition-metal complexes. In order to study the effect upon its coordination properties of the substitution of a phenyl for a methyl group in the acetylacetonate ion, and as part of a general investigation of spectroscopic pseudosymmetry of low-symmetry complexes, we have measured the EPR spectrum of ~ 0.5 mole % Cu^{2+} in palladium bis-benzoylacetonate, $Pd(benzac)_2$, the molecular structure of which is shown in Fig. 2. This chapter reports some of the features of this investigation and also describes a useful least-squares method which we employ to determine the principal g tensor axes and components.

*Supported by the Advanced Research Projects Agency Contract SD-131 through the Materials Research Laboratory at the University of Illinois, and the Illinois Graduate Research Board.

The method generally used for arriving at g values is due to Schonland.[1] For this one measures g^2 for rotations of the magnetic vector in three crystal planes, calculates from this the g tensor in terms of an arbitrary crystal coordinate system, and then diagonalizes the tensor to find principal g values. Our procedure differs somewhat from Schonland's method. We express the magnetic field vector in terms of an arbitrarily defined *molecular* coordinate system for each measurement and then calculate and diagonalize the g tensor in this axis system to obtain principal g values and the associated Euler angles relating them to the molecular axes. Although the details have only minor novelty, the method will be described here because we have found it extremely convenient. We have programmed it so that the computer accepts the crystallographic parameters, rotation data, and g^2 measurements and carries out a least-squares analysis. The fact that this method directly relates the principal g axes to the molecular orientation is particularly useful when, as with Cu(benzac)$_2$, the effective ligand-field symmetry is not readily apparent and is one of the principal points in question.

PROCEDURE FOR COMPUTING PRINCIPAL g VALUES

Values of g^2 are measured for rotations in several crystal planes. For each measurement the direction of the magnetic field is defined in terms of the monoclinic crystal coordinates **a**, **b**, **c**, and $\beta^*(\beta^* = 180° - \beta)$ by means

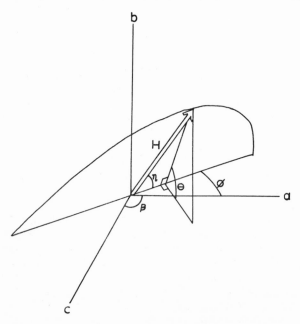

Fig. 1. Diagram showing the angles defining the magnetic vector **H** in terms of a monoclinic crystal coordinate system.

of the angles η, θ, and ϕ (Fig. 1). Here θ is the angle made by the plane of rotation with the **ac** plane. If the plane of rotation cuts the **ac** plane along a line **l**, then ϕ is the angle which **l** makes with the **a** axis and η is the angle which the magnetic vector **H** makes with **l**. Thus θ and ϕ are constants for any rotation, and may be calculated from the morphology of the crystal, while η is the angle varied during the rotation. The value of measuring the magnetic field direction in this way is that for a monoclinic space group with a single molecule in the asymmetrical unit the magnetic vector makes equivalent projections on the different molecules in the unit cell when it lies in the **ac** plane (or along **b**). Thus in general two sets of lines are observed for a monoclinic crystal, but these coalesce into a single set when **H** lies in the **ac** plane, so that this provides a good internal reference point from which to measure the position of **H**. Similar relationships hold for orthorhombic space groups.

The expressions for a unit magnetic field vector in terms of the crystallographic axes are:

$$H_a = H_c \cos \beta^* \pm (1 - H_c^2 \sin^2 \beta^* - \sin^2 \eta \sin^2 \theta)^{1/2}$$

[the sign being decided by which of the quadrants defined by the (100) and (001) planes **H** lies in]

$$H_b = \sin \eta \sin \theta$$

and

$$H_c = \cos \phi (\sin \eta \cos \theta - \cos \eta \tan \phi)/\sin \beta^*$$

These are then converted to an orthogonal crystal coordinate system by the relationships

$$H_a' = H_a \sin \beta^*, \qquad H_b' = H_b, \qquad H_c' = H_c - H_a \cos \beta^*$$

A molecular coordinate system is now defined. This is quite arbitrary and is just a convenient orientation to which the EPR measurements may be referred. That chosen for Pd(banzac)$_2$ is shown in Fig. 2. Multiplication by a transformation matrix then gives **H** in terms of these molecular axes, and for a monoclinic space group, multiplication by a symmetry-related matrix gives the molecular projections for the second nonequivalent molecule in the unit cell:

$$\begin{bmatrix} H_{x1} \\ H_{y1} \\ H_{z1} \end{bmatrix} = \begin{bmatrix} a_1 & b_1 & c_1 \\ a_2 & b_2 & c_2 \\ a_3 & b_3 & c_3 \end{bmatrix} \begin{bmatrix} H_a' \\ H_b' \\ H_c' \end{bmatrix}; \qquad \begin{bmatrix} H_{x2} \\ H_{y2} \\ H_{z2} \end{bmatrix} = \begin{bmatrix} a_1 & -b_1 & c_1 \\ a_2 & -b_2 & c_2 \\ a_3 & -b_3 & c_3 \end{bmatrix} \begin{bmatrix} H_a' \\ H_b' \\ H_c' \end{bmatrix}$$

In the present problem the g value differing most from 2 was associated with the projection having the larger value of H_z.

To first order, which is a good approximation at the field used here ($\sim 11,500$ G), the measured g values may be fitted to the expression

$$g^2 = H_x^2 g_{xx}^2 + H_y^2 g_{yy}^2 + H_z^2 g_{zz}^2 + 2H_x H_y g_{xy}^2 + 2H_x H_z g_{xz}^2 + 2H_y H_z g_{yz}^2$$

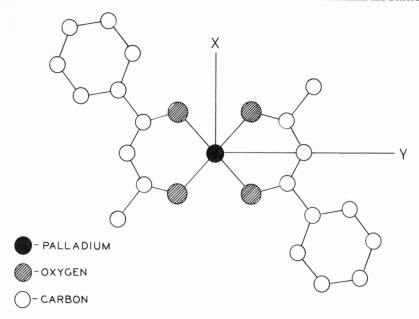

- PALLADIUM

- OXYGEN

- CARBON

Fig. 2. The axis system used to define the Pd(benzac)$_2$ molecule in the analysis of the EPR measurements. The z axis is perpendicular to the plane of the molecule.

The set of n simultaneous equations for the n measurements of g^2 is solved to give a best least-squares fit for g_{ij}^2, these being the elements of the g tensor expressed in the molecular coordinate system. The matrix g_{ij}^2 is then diagonalized by a unitary transformation to give the principal g values.

The elements of the transformation matrix T_{ij} can be used to obtain the Euler angles by which the molecular axes must be rotated to convert them into the principal g axes. The Euler angles are defined as follows. First, rotate by A counterclockwise about **z** to give **x′**, **y′**, **z**, then by B about **x′** to give **x′**, **y″**, **z′**, then by C about **z′** to give **x″**, **y‴**, **z′**. These relationships easily follow:*

$$\cos B = T_{33}; \qquad \cos(C + A) = (T_{11} + T_{22})/(1 + T_{33});$$

$$\cos(C - A) = (T_{11} - T_{22})/(1 - T_{33})$$

It should be noted that either T_{ij} or $-T_{ij}$ will diagonalize g_{ij}^2. Only one of these matrices, however, produces internally consistent values for the Euler angles.

In order to diagonalize g^2, one must use at least three sets of results corresponding to sections of the g ellipsoid. In general, two sets of results are obtained for each rotation. However, these sets are not independent for any rotation containing the **b** axis.

*For the expression of the elements T_{ij} in terms of Euler angles see Goldstein.[2]

The same method can be used to treat the values of K^2g^2 to obtain the principal values of the hyperfine tensor [provided that the matrix diagonalizing $(Kg)^2_{ij}$ is not very different from that diagonalizing g^2_{ij}].[3] The procedure described here is carried out using a FORTRAN II computer program. As well as calculating the principal g values and Euler angles, the molecular projections of the H vector are listed for each measurement together with the root-mean-square error in g^2. The total computation time to process 150 values each of g^2 and g^2K^2 is about 15 sec on the IBM 7094 computer. The program has also been modified to correct the g values for second-order hyperfine effects.

EXPERIMENTAL

Crystals of $Cu/Pd(benzac)_2$ were grown by allowing ethanol to diffuse slowly into a chloroform solution of the palladium complex containing $\frac{1}{2}$ mole % of the copper compound. The crystals were formed as long needles (α form) and small diamonds (β form), the principal faces of which were determined by means of an x-ray precession camera and optical goniometry. The EPR spectra were measured by placing a crystal upon a known face on the quartz rod and screwing this into the cavity of a Varian spectrometer operating at 35 GHz. The direction of the magnetic field was varied by rotating the magnet.

INTERPRETATION OF THE SPECTRA

$Pd(benzac)_2$ crystallizes in the monoclinic space group $P2_{1/c}$ with unit cell parameters $a = 9.367$ Å, $b = 10.518$ Å, $c = 9.454$ Å, and $\beta = 108°$.[4] Although the palladium and copper compounds are not isomorphous, the molecular structures are virtually identical.[5]

The best-developed face on β-$Pd(benzac)_2$ is (100), while (11$\bar{1}$), etc., are also well developed. Rotations were made in each of these planes, and, as expected, two sets of four lines each appeared owing to the interaction of the unpaired electron with the copper nuclear spin $I = \frac{3}{2}$ for each of the two molecules in the unit cell. As the value of K (the measured hyperfine splitting) increased, first the two outer lines, then all four hyperfine lines, split into two components (Fig. 3a) due to the slightly different nuclear magnetic moments of the two naturally occurring isotopes ^{63}Cu (69.09%) and ^{65}Cu (30.91%), both of which have spin $I = \frac{3}{2}$. The measured ratio of the hyperfine coupling constants, $K(^{63}Cu)/K(^{65}Cu) \approx 0.937$, agreed reasonably well with the ratio of the nuclear magnetic moments, $g_N(^{63}Cu)/g_N(^{65}Cu) = 0.9337$.

Forbidden Transitions

When H lay near the molecular xy plane several additional lines were observed between the three high-field allowed EPR lines (Fig. 3b). These extra lines correspond to the formally forbidden transitions $\Delta m_S = \pm 1$, $\Delta m_I = \pm 1$ which gain intensity by nuclear Zeeman and nuclear quadrupole

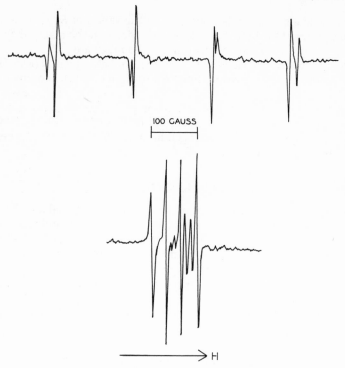

Fig. 3. The EPR spectrum of Cu(benzac)$_2$ when **H** lies (a) near the **z** molecular
axis, (b) near the **xy** molecular plane.

mechanisms.* Six such transitions are possible, with spacings and intensities
dependent upon the nuclear-Zeeman, hyperfine, and nuclear-quadrupole
interactions. We are examining these "forbidden" transitions at other
magnetic field strengths to fix assignments and determine the constants of
the effective spin Hamiltonian, and a more detailed account will be forth-
coming.† Thus far we have observed only the four $\Delta m_I = \pm 1$ transitions of
lowest frequency.

Angular Variation of Lines

The needle axis of α-Pd(benzac)$_2$ was the (101) direction, with the well-
developed faces being (11$\bar{1}$), etc. The rotation about the needle axis [in the
(101) plane] showed the expected two sets of lines, coalescing into one set
when **H** lay in the **ac** plane or along **b**. However, the rotation in the (11$\bar{1}$)
plane showed two pairs of sets of lines, one pair having a lower intensity than
the other pair, and with the ratio of the intensities between the pairs of sets

*Choh and Seidel[6] have recently discussed a somewhat similar case in which the nuclear
Zeeman mechanism dominates.
†Note added in proof. So and Belford interpret these lines in detail and obtain accurate eqQ
values, as summarized in a preliminary communication.[16]

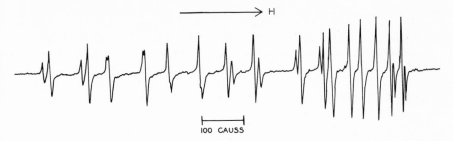

Fig. 4. The EPR spectrum of Cu^{2+} in a typical crystal of α-Pd(benzac)$_2$ with the magnetic field rotating in the (11$\bar{1}$) crystallographic plane.

varying somewhat from crystal to crystal. A typical spectrum is shown in Fig. 4. These lines coalesced into one set of lines when **H** was in the **ac** plane and two sets of lines when $\eta = \pm 90°$. The variation of g^2 with angle for the (11$\bar{1}$) faces of the α and β forms of Pd(benzac)$_2$ is shown in Fig. 5.

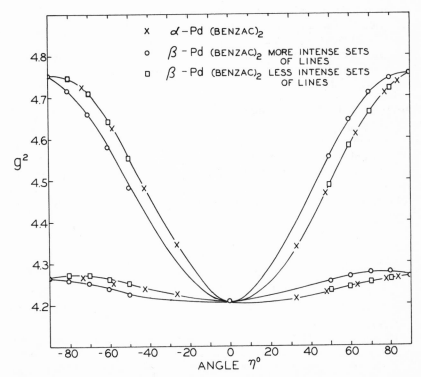

Fig. 5. The variation of g^2 with angle for rotations of the magnetic vector in the (11$\bar{1}$) planes of α- and β-Pd(benzac)$_2$.

Twinning

It is apparent that the two pairs of sets of lines observed for α-Pd(benzac)$_2$ in the $(11\bar{1})$ rotation correspond to a superposition of the spectra observed for β-Pd(benzac)$_2$ when η varies from $0°$ to $+90°$ and from $0°$ to $-90°$ in the $(11\bar{1})$ rotation. This can be explained if the α form of Pd(benzac)$_2$ is twinned along a plane containing the needle axis, this plane being $(10\bar{1})$ for one twin and $(\bar{1}01)$ for the other twin. To test this, we ground down the edges of several crystals along a direction normal to the $(10\bar{1})$ and $(\bar{1}01)$ planes of the two twins. In each case the intensity of one of the pairs of sets of lines decreased markedly. Thus we conclude that twinning of the kind described occurs in the α form of Pd(benzac)$_2$. It should be noted that only two sets of lines could be observed in the rotation about the needle axis because of the near equality of the a and c unit-cell dimensions. For the same reason the twinning is not readily apparent from the x-ray precession pictures,[4] so that in this case EPR gives useful information on the crystal morphology which is not easily obtained by the usual means.

The g Values and Hyperfine Constants

The principal g values of Cu^{2+} diluted into the two forms of Pd(benzac)$_2$ were calculated by the method described. The results are listed in Table I. As the $(11\bar{1})$ rotation yielded two independent sets of values of g^2 (as this plane does not contain the b axis), four sets of data could be used in the calculations, and combining these in different ways provided a check on the consistency of the results.

Anisotropy and Effective Symmetry

The final values $g_x = 2.049_0$, $g_y = 2.047_8$, and $g_z = 2.242_8$ obtained after correcting for second-order hyperfine effects show that g_x and g_y are almost equal, with g_x slightly, but probably just significantly, greater than g_y. We believe that the difference is real, with $g_x > g_y$, partly because the same answer is obtained from every pair of rotations. The Euler angles show that within experimental error the principal g axes lie along the axes defined for the molecule.

TABLE I
The Principal g Values and Euler Angles Relating the Molecular Axes to the Principal g Axes for Cu(benzac)$_2$

Planes of rotation	Rms deviation	Principal g values			Euler angles[a] (deg)		
		g_x	g_y	g_z	A	B	C
(101), $(11\bar{1})$, (100)	0.001	2.0489	2.0477	2.2429	1.6	1.4	-1.8
$(11\bar{1})$, (101)	0.001	2.0491	2.0480	2.2429	1.3	1.1	-12.6
$(11\bar{1})$, (100)	0.001	2.0491	2.0476	2.2426	1.6	-1.4	3.6

[a]For the definition of the Euler angles see text.

Thus one sees that the effective ligand field symmetry in $Cu(benzac)_2$ is D_{2h}, but very close to D_{4h}. The fact that the in-plane g axes lie, within experimental error, along the $O-Cu-O$ angle bisectors (or at least much closer to the bisectors than to the $Cu-O$ bonds) suggests that as far as the perturbation on the d orbitals is concerned, the two oxygens on each benzac ligand are equivalent. If this were not the case, the effective point group would be C_{2h} (Fig. 2), in which d_{xz} and d_{yz} belong to the same representation. These orbitals could then mix to give two new orbitals whose directional properties would be a function of the C_{2h} perturbation on the D_{2h} point group, and this would cause a corresponding rotation of the in-plane g axes. This point is discussed in detail elsewhere.[7] A large rotation of this kind has in fact been observed in Cu(II) bis-N-methylsalicylaldimine[7] diluted into the corresponding nickel compound. The apparent equivalence of the two oxygen atoms in the chelated benzac ligand is perhaps somewhat surprising, as in both $Cu(benzac)_2$ and $Pd(benzac)_2$ the metal–oxygen bond cis to the benzene ring is 0.01–0.02 Å shorter than that $trans$ to the benzene ring.

The slight difference between g_x and g_y could be caused by a small energy difference ($\sim 500 \, \text{cm}^{-1}$) between the d_{xz} and d_{yz} orbitals and/or by a slight difference between their orbital coefficients. A very small admixture (caused by the C_{2h} perturbation) of the d_{3z-r^2} orbital into d_{xy} is possible, but can be shown to be of negligible effect on g_x and g_y providing that the g principal axes lie close to the bond-angle bisectors, as we shall demonstrate in a separate note. It may be noted that Maki and McGarvey observed[8] a slight rhombic distortion ($g_x = 2.0551$, $g_y = 2.0519$) in the in-plane g values of Cu(II) bis-acetylacetonate [hereafter referred to as $Cu(acac)_2$], though they considered this difference to be within the limits of their experimental error.

Hyperfine Parameters

While in principle values of $K^2 g^2$ can be treated in a manner analogous to g^2 to yield the principal hyperfine parameters, we found in practice that the great difference in magnitude between $A^2 g_{\parallel}^2$ and $B^2 g_{\perp}^2$ (50:1) caused rather large standard deviations ($\sim 10\%$) in the calculation of the in-plane hyperfine values by this method. In particular, the overlap of hyperfine lines due to the two copper nuclei should probably be considered, and the computer program is being modified to take this effect into account. The values of the hyperfine constants used in the calculations which follow will thus be those measured along the x, y, and z directions, these being

$$A_z = -181 \times 10^{-4} \, \text{cm}^{-1}, \qquad A_x \approx A_y = B = -28 \times 10^{-4} \, \text{cm}^{-1}$$

Calculation of Bonding Parameters

The very close similarity of g_x to g_y in $Cu(benzac)_2$ and the fact that the electronic spectrum shows only one band assignable to the transitions from the d_{xz} and d_{yz} orbitals to the ground-state orbital suggest that to a good approximation the bonding parameters of the molecule may be calculated using the point group D_{4h}.

The molecular orbitals for a copper ion surrounded by a square planar arrangement of oxygen atoms are, in the probable order of increasing energy,

$$E_g = \beta d_{xz} - (1/\sqrt{2})\beta'(p_z^{(1)} - p_z^{(3)})$$

$$= \beta d_{yz} - (1/\sqrt{2})\beta'(p_z^{(2)} - p_z^{(4)})$$

$$B_{2g} = \gamma d_{xy} - \tfrac{1}{2}\gamma'(p_y^{(1)} + p_x^{(2)} - p_y^{(3)} - p_x^{(4)})$$

$$A_{1g} = \eta d_{3z^2 - r^2} - \tfrac{1}{2}\eta'(\sigma_x^{(1)} + \sigma_y^{(2)} - \sigma_x^{(3)} - \sigma_y^{(4)})$$

$$B_{1g} = \alpha d_{x^2 - y^2} - \tfrac{1}{2}\alpha'(- \sigma_x^{(1)} + \sigma_y^{(2)} + \sigma_x^{(3)} - \sigma_y^{(4)})$$

It should be noted that in the point group D_{4h} as it is ordinarily used the x and y axes are defined as directed toward the oxygen atoms, and the ground state for the hole is the $d_{x^2 - y^2}$ orbital. The ligand atoms are assumed to use 2s and 2p orbitals to bond with the copper 3d orbitals, and the σ orbitals are defined by

$$\sigma^{(i)} = np^{(i)} \pm (1 - n^2)^{1/2}s^{(i)}$$

With neglect of nuclear Zeeman and quadrupole effects the perturbing part of the Hamiltonian for a d^9 ion under the influence of spin-orbit coupling and an applied magnetic field $\mathbf{H}$ is[9]

$$W' = \lambda(r)\mathbf{L} \cdot \mathbf{S} + \beta_0\mathbf{H} \cdot (\mathbf{L} + 2.0023\mathbf{S}) + 2g_N\beta_N\beta_0$$

$$\times \{[(\mathbf{L} - \mathbf{S}) \cdot \mathbf{I}/r^3] + [3(\mathbf{r} \cdot \mathbf{S})(\mathbf{r} \cdot \mathbf{I})/r^5] - (8\pi/3)\delta(r)\mathbf{S} \cdot \mathbf{I}\}$$

where the symbols have their usual significance.

On applying this Hamiltonian to the above wave functions, to second order, assuming that B_{1g} is the ground state, one obtains the following spin Hamiltonian:

$$\mathscr{H} = \beta_0[g_{\parallel}\mathscr{H}_z S_z + g_{\perp}(H_x S_x + H_y S_y) + A I_z S_z + B(I_x S_x + I_y S_y)]$$

The g values and hyperfine constants in this expression are*

$$\Delta g_{\parallel} = -8e(\alpha\gamma) + C_1 \tag{1a}$$

$$\Delta g_{\perp} = -2\mu(\alpha\beta) + C_2 \tag{1b}$$

$$A = P[-\tfrac{4}{7}\alpha^2 - K + \Delta g_{\parallel} + \tfrac{3}{7}\Delta g_{\perp} - C_1 - \tfrac{3}{7}C_2] \tag{1c}$$

$$B = P[\tfrac{2}{7}\alpha^2 - K + \tfrac{11}{14}\Delta g_{\perp} - \tfrac{11}{14}C_2] \tag{1d}$$

where $\Delta g_{\parallel} = g_{\parallel} - 2.0023$, $\Delta g_{\perp} = g_{\perp} - 2.0023$, $e = \lambda_0\alpha\gamma/(E_{x^2 - y^2} - E_{xy})$, $\mu = \lambda_0\alpha\beta/(E_{x^2 - y^2} - E_{xz,yz})$, λ_0 is the free-ion spin-orbit coupling constant, P is the free-ion dipole term given by $2g_N\beta_n\langle r^{-3}\rangle$, and K is the isotropic hyperfine constant in the complex. The parameters C_1 and C_2 are corrections which should be added to take into account the unpaired electron density

*The expressions derived by Kivelson and Neiman[10] have been used, except that K_0^2 has been replaced by K.

associated with the ligands, and may be estimated as

$$C_1 = 8e[\alpha'\gamma S + \tfrac{1}{2}\alpha'(1 - \gamma^2)^{1/2}T(r)]$$

$$C_2 = 2\mu[\alpha'\beta S + (1/\sqrt{2})\alpha'(1 - \beta^2)^{1/2}T(r)]$$

In this approximation all overlaps between ligand and metal orbitals have been neglected except that between $d_{x^2-y^2}$ and the ligand σ orbitals, where the coefficients are given by $\alpha^2 + \alpha'^2 - 2\alpha\alpha'S = 1$. The $T(r)$ involves an integration over ligand s and p orbitals.

In principle, it is possible to solve Eq. (1a)–(1d) simultaneously to obtain values for the orbital coefficients. In practice, however, the equations are probably better solved by an iterative procedure. The parameters C_1 and C_2 may be set equal to zero and values of α, β, γ, and K easily calculated. These may then be used to calculate C_1 and C_2, which may be applied as corrections to $\Delta g_{\parallel}$ and $\Delta g_{\perp}$. New values may then be calculated for the orbital coefficients, and the process repeated to the desired degree of accuracy.

The values used for $T(r)$, P, and S were those estimated by Kivelson and Neiman:[10] $T(r) = 0.22$, $P = 0.036 \, \text{cm}^{-1}$, $S = 0.076$, while the spectroscopically determined value of $-828 \, \text{cm}^{-1}$ was used for λ_0. The estimates of $E_{x^2-y^2} - E_{xy}$ and $E_{x^2-y^2} - E_{xz,yz}$ may be obtained from the optical spectrum of Cu(benzac)$_2$ (although it should be noted that this assumes an identical ordering of the energy levels in the pure copper compound and in the diluted species). The electronic spectrum of pure Cu(benzac)$_2$ consists of three bands[11] at 14,500, 16,100, and 18,100 cm^{-1}, and the temperature dependence of the spectrum suggests that the band at 18,100 cm^{-1} may be fairly unequivocally assigned to the transitions $d_{xz}, d_{yz} \rightarrow d_{x^2-y^2}$. Although the optical spectrum does not at present allow a definite assignment to be made for the other two bands, a molecular orbital calculation[12] on the similar compound bis(dipivaloylmethanido) copper(II) suggested the ordering $E(d_{3z^2-r^2}) > E(d_{xy})$, which would place $(E_{x^2-y^2} - E_{xy})$ at 16,100 cm^{-1} for Cu(benzac)$_2$.

TABLE II
The Orbital Coefficients and Isotropic Hyperfine Parameters for Cu(benzac)$_2$ and Cu(acac)$_2$

Compound	Orbital coefficients[a] without ligand corrections				Orbital coefficients with ligand corrections			
	α^2	β^2	γ^2	K^2	α^2	β^2	γ^2	K^2
Cu(benzac)$_2$	0.756	0.665	0.773	0.330	0.779	0.722	0.816	0.339
Cu(acac)$_2$ [b]	0.742	0.77	0.875	0.306	0.763	0.828	0.915	0.317

[a]See text for the definition and method of calculation of these parameters.
[b]Calculated using the measured values reported by Maki and McGarvey.[8]

The results of the calculations of the orbital parameters for Cu(benzac)$_2$ are shown in Table II. The use of $E_{x^2-y^2} - E_{xy} = 14,500$ cm^{-1} would change the final calculated value of γ^2 from 0.816 to 0.735, making it very similar to β^2. Although such a value is certainly possible, it does seem perhaps more reasonable that the in-plane π bonding should be less covalent than the out-of-plane π bonding. It is apparent that the ligand correction terms C_1 and C_2 make a significant ($\sim 8\%$) change in the calculated coefficients, particularly in β^2 and γ^2. The fact that $\beta^2 > \alpha^2$ for Cu(benzac)$_2$ is perhaps surprising, as it suggests that more delocalization occurs in the out-of-plane π bonding than in the σ bonding. In fact, the approximations made in the calculations certainly do not warrant such a conclusion; all it seems safe to conclude is that there is considerable π bonding in the complex.

Comparison of the Bonding Parameters with Cu(acac)$_2$

Maki and McGarvey[8] observed the values $g_{\parallel} = 2.266$, $g_{\perp} = 2.0535$, $A = -1.6 \times 10^{-2}$ cm^{-1} and $B = -0.195 \times 10^{-2}$ cm^{-1} for Cu^{2+} diluted into Pd(acac)$_2$. Using a method analogous to that reported here, except for the neglect of the overlap terms $\alpha'\gamma S$ and $\alpha'\beta S$, these authors calculated orbital coefficients of $\alpha^2 = 0.81$, $\beta^2 = 0.99$, and $\gamma^2 = 0.85$ from these results. However, the interpretation of the electronic spectrum of Cu(acac)$_2$ was at that time uncertain, and the assignments $(E_{x^2-y^2} - E_{xz,yz}) = 25,000$ cm^{-1} and $(E_{x^2-y^2} - E_{xy}) = 15,000$ cm^{-1} were used in the calculations. More recent work[13] suggests that the correct assignments are $(E_{x^2-y^2} - E_{xz,yz}) = 18,500$ cm^{-1} and $(E_{x^2-y^2} - E_{xy}) = 16,300$ cm^{-1}. Consequently, the orbital coefficients have been recalculated using Eqs. (1a)–(1d) and the results are given in Table II. It is clear that the new assignments produce a much more reasonable value for the coefficient β^2. Comparison of the values calculated for Cu(acac)$_2$ with those for Cu(benzac)$_2$ shows that while the σ bonding coefficients are almost the same, the two π-bonding coefficients are both ~ 0.1 lower in the latter complex. It is tempting to ascribe this to a mesomeric effect involving the phenyl group, and crystallographic data provide some evidence that conjugation of the benzene ring with the rest of the molecule is important in complexes of the benzoylacetonate anion. In each of the four complexes of this ligand whose structures have been determined by x-ray techniques* the benzene rings have been approximately coplanar with the rest of the ligand.

ACKNOWLEDGMENTS

We acknowledge Research Support from the Advanced Research Projects Agency Contract SD-131 through the Materials Research Laboratory at the University of Illinois, and the University of Illinois Graduate Research Board.

*See Hon et al.[4,5] for the complexes discussed here, plus Hon et al.[14] for VO(benzac)$_2$ and Belford et al.[15] for Zn(benzac)$_2 \cdot$ C$_2$H$_5$OH.

REFERENCES

1. D. S. Schonland, *Proc. Phys. Soc.* (*London*) **73**, 788 (1959).
2. H. Goldstein, *Classical Mechanics* (Addison-Wesley, Reading,˙ Massachusetts, 1950), p. 107.
3. A. Lund and T. Vänngård, *J. Chem. Phys.* **42**, 2979 (1965).
4. P. K. Hon, C. E. Pfluger, and R. L. Belford, *Inorg. Chem.* **6**, 730 (1967).
5. P. K. Hon, C. E. Pfluger, and R. L. Belford, *Inorg. Chem.* **5**, 516 (1966).
6. S. H. Choh and G. Seidel, *Phys. Rev.* **164**, 412 (1967).
7. B. W. Moore, Ph.D. Thesis, University of Illinois, Urbana, Illinois, 1968; presented in part by R. L. Belford at the Symposium on Electron Spin Resonance, and by B. W. Moores and R. L. Belford at the National Meeting of the American Chemical Society, San Francisco, California, April 2, 1968. Cf. M. A. Hitchman, C. D. Olson, and R. L. Belford, *J. Chem. Phys.* **50** (1969).
8. A. H. Maki and B. R. McGarvey, *J. Chem. Phys.* **29**, 31 (1958).
9. A. Abragam and M. H. L. Pryce, *Proc. Roy. Soc.* (*London*) **A205**, 135 (1951).
10. D. Kivelson and R. Neiman, *J. Chem. Phys.* **35**, 149 (1961).
11. R. L. Belford and M. A. Hitchman, Single-crystal spectra, to be published.
12. F. A. Cotton, C. B. Harris, and J. J. Wise, *Inorg. Chem.* **6**, 909 (1967).
13. J. Ferguson, R. L. Belford, and T. S. Piper, *J. Chem. Phys.* **37**, 1569 (1962); T. S. Piper and R. L. Belford, *Mol. Phys.* **5**, 169 (1962).
14. P. K. Hon, R. L. Belford, and C. E. Pfluger, *J. Chem. Phys.* **43**, 1323 (1965).
15. R. L. Belford, N. D. Chasteen, M. A. Hitchman, P. K. Hon, C. E. Pfluger, and I. C. Paul, *Inorg. Chem.* **8** (1969).
16. H. So and R. L. Belford, *J. Am. Chem. Soc.* **91** (1969).

Spectral Properties of Oxovanadium(IV) Complexes. IV. Correlation of ESR Spectra with Ligand Type

L. J. Boucher, Edmund C. Tynan, and Teh Fu Yen

Department of Chemistry
Mellon Institute, Carnegie-Mellon University
Pittsburgh, Pennsylvania

ESR spectral parameters, A_0, $A_\parallel$, $A_\perp$ and g_0, $g_\parallel$, $g_\perp$ as well as electronic absorption maxima are listed for a number of oxovanadium(IV) complexes with a series of representative ligands. The four donor atoms of the ligands are: oxygen (β-diketones, oxalate, water), oxygen–nitrogen (β-ketimines, salicylaldimines), nitrogen (porphyrins, thiocyanate), sulfur (dithiocarbonate), chloride, and cyanide. The electronic structures of the square-based pyramidal oxovanadium(IV) complexes are discussed with the aid of spectral data. A plot of g values versus A values for each of the complexes maps out regions for donor types. Within each group, solvent and substituent effects on the spectra are also noted. The origin of the ESR correlation is related to the empirical dependence of the isotropic contact term on the molecular orbital coefficients and energy of the vanadium orbitals. Extension of this approach to systems in which the donor atom is unknown is suggested.

INTRODUCTION

Studies of electron spin resonance spectra of paramagnetic transition-metal chelates can lead to a detailed description of the electronic structure of these compounds.[1] In a more qualitative way ESR spectroscopy is also useful in solving problems related to the binding of metals to organic ligands. This is particularly true when the method is applied to metal chelates in complex biological systems. Ideally, ESR spectroscopy can lead here to a number of kinds of information: (1) identification of the metal, its oxidation state, and its spin state; (2) identification of the binding site (ligands) and its symmetry; and (3) determination of the concentration of the paramagnetic metal ion. A further experimental advantage is the fact that ESR measurements are several orders of magnitude more sensitive than most spectroscopic techniques and require relatively small samples. In addition, spectra can easily be determined for solids and solutions even when the metal chelate is a minor component in a complex mixture, since most materials are diamagnetic.

The methodology of ESR spectroscopy has been quite successfully applied to naturally occurring Cu^{2+} and Fe^{2+} chelates.[2] Unfortunately, other paramagnetic transition-metal complexes in biological systems have not received much attention. Vanadium is an important trace element in nature.[3] For example, it is concentrated by the marine animal, *Phallusia mamillata* (sea squirt),[4] and land plant, *Amanita muscaria* (mushroom).[5] Although vanadium-containing fractions have been isolated from these systems, the exact binding of the vanadium atom is uncertain. Vanadium is also widely concentrated in petroleum (0–6000 ppm).[6] To a very large extent the vanadium is contained in the solid asphaltic fraction.[7] A part (approximately 50%) of the vanadium compounds is easily extracted with organic solvents. These materials have been chemically characterized as porphyrin complexes.[8] The remaining vanadium, which is not easily removed from the asphaltene, is classified as nonporphyrin vanadium. Unfortunately, the nonporphyrin complexes have not been individually isolated and the binding of the metal is unknown.

ESR spectroscopy is ideally suited to the study of vanadium in natural systems. The vanadium-51 nucleus (approximately 100% abundant) has a large nuclear moment and a nuclear spin of $\frac{7}{2}$. This gives rise to an easily-resolvable characteristic eight-line spectrum. Common valence states of vanadium are $+2$, $+3$, $+4$, $+5$. The $+5$ state is diamagnetic and the $+3$ state, although paramagnetic, is usually not observable by ESR due to internal electric field effects. Both the $+2$ and the $+4$ states can, however, be detected at room temperature. Vanadium $+2$ is oxidatively unstable and easily yields the oxycation, VO^{2+}. The vanadium $+4$ oxidation state exists almost exclusively as the oxovanadium(IV) ion in organic complexes. Furthermore, the $+4$ state requires a noncubic field for ESR observability, and this is fulfilled by the oxovanadium(IV)-type complexes. Another simplifying feature of the materials is the absence of interelectron interaction effects, since there is only one d electron. In this chapter it is assumed that all naturally occurring paramagnetic vanadium exists as complexes of the VO^{2+} cation.

In order to use ESR spectroscopy to obtain information about the binding of vanadium in biological systems, suitable data on model compounds must exist. Due to the current interest in the electronic structure of VO^{2+} complexes[9] there are for the first time sufficient data in the literature for a wide variety of ligands. In this chapter we shall attempt to obtain a correlation between the ESR parameters and the ligand type in oxovanadium-(IV) complexes. We shall then show the basis of this correlation in the electronic structure of the complexes and the theory of ESR spectra of transition-metal complexes. Finally, we shall show how the ligand correlation can be applied to the case of a naturally occurring vanadium complex in petroleum.

ELECTRONIC STRUCTURE OF OXOVANADIUM(IV) COMPLEXES

In order to discuss the ESR spectra of VO^{2+} complexes, certain details of their molecular and electronic structure must be at hand. The crystal and molecular structure of a number of VO^{2+} complexes have been reported. The data obtained for bis-acetylacetonatooxovanadium(IV), $VO(acac)_2$,

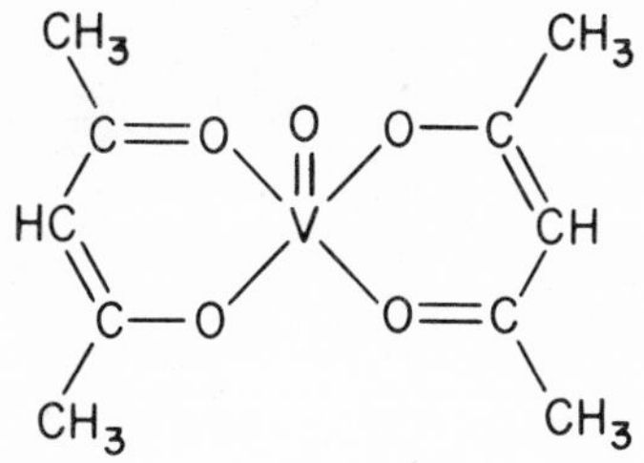

$$VO(acac)_2$$

Fig. 1. Structural representation for
bis-acetylacetonatooxovanadium(IV).

typify these results[10] (Fig. 1). The fivefold coordination about the metal is square pyramidal. The axial oxygen atom is significantly closer to the vanadium ion than are the four β-diketone oxygens. The four diketone oxygen atoms form a plane which is the base of the pyramid. Finally, the vanadium atom is considerably above this plane. It is assumed that the vast majority of oxovanadium complexes will maintain the general structural features of the $VO(acac)_2$. Of course, the $V=O$ distance, vanadium-to-ligand, and the vanadium-to-plane distances should vary in a significant way from complex to complex.

The overall molecular symmetry for $VO(acac)_2$ is C_{2v}. If only the donor atoms are considered, then the symmetry is raised to C_{4v}. Even for lower-symmetry cases the effective symmetry appears to be C_{4v}, so the labels and discussion appropriate to this symmetry will be given here. It is important to note that the VO^{2+} complexes can also be six coordinate. The five-coordinate complexes weakly add a sixth ligand to the vanadium atom in the axial coordination position *trans* to the oxygen atom. Since the vanadium-to-ligand bond length is substantially longer in this case than the in-plane ligands,[11] the interaction is justifiably considered only a perturbation on the square pyramidal system.

Ballhausen and Gray[12] first formulated a ligand field picture of oxo-vanadium(IV) complexes. For the C_{4v} symmetry, d^1 case, the d-energy level ordering is

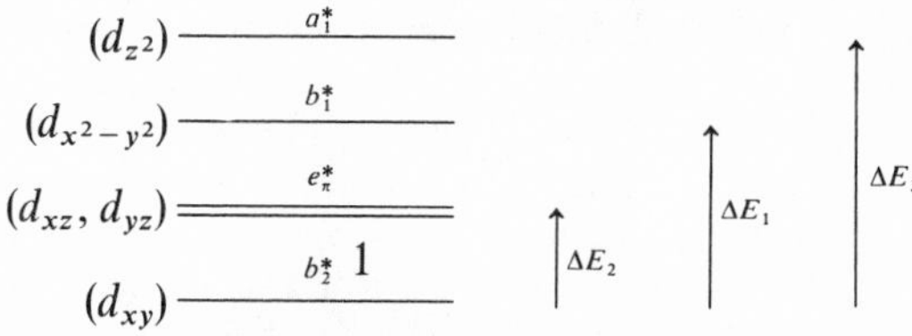

$$(d_{z^2}) \; \underline{\quad a_1^* \quad}$$
$$(d_{x^2-y^2}) \; \underline{\quad b_1^* \quad}$$
$$(d_{xz}, d_{yz}) \; \underline{\overline{\quad e_\pi^* \quad}}$$
$$(d_{xy}) \; \underline{\quad b_2^* \; 1 \quad}$$

$\Delta E_2 \quad \Delta E_1 \quad \Delta E_3$

The original pure metal d-orbital designations and their respective orbital coefficients are given in the diagram. In the complexes the unpaired electron is located in the d_{xy} orbital. The three expected low-energy ligand-

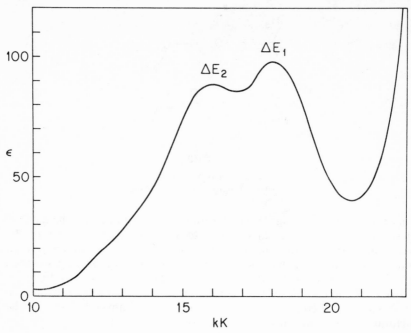

Fig. 2. Electronic absorption spectrum of bis-acetylacetonateethylenediimine-oxovanadium(IV) in ethanol.

field bands are shown. A typical absorption spectrum of a VO^{2+} complex is given in Fig. 2. The assigned ΔE_1 and ΔE_2 transitions are labeled in the spectrum. The ΔE_3 transition is usually of higher energy and often obscured by intense low-lying charge-transfer and interligand transitions.[13] In general, there is some argument about the assignment of the electronic transition for the complexes. The assignments of ΔE_1 and ΔE_2 given here seem plausible and agree with most current evidence. In arriving at the energy-level scheme, the $d_{x^2-y^2}$ level is considered a σ-antibonding orbital interacting with the in-plane ligands; the d_{z^2} is σ-antibonding with the axial oxygen atom; the d_{xz}, d_{yz} level is considered to be π-antibonding with the axial oxygen atom; and, finally, the d_{xy} orbital is nonbonding or is π-antibonding with the in-plane ligands. The positions of the absorption maxima for a number of oxovanadium complexes are given in Table I. The frequency of ΔE_3 is not given, since its actual assignment is in doubt. In addition, it is not needed for the following discussion. The complexes listed in Table I were selected because complete ESR data are available for all of them. Furthermore, all the materials are well-defined complexes of C_{2v} or C_{4v} microsymmetry. Fortunately, a wide range of ligand types is represented.

The d–d transitions are not listed for two of the complexes. The bands are obscured by intense ligand and charge-transfer bands in the spectra.

TABLE I
Visible Absorption Maxima of Oxovanadium Complexes

Complex[a]	Host	ΔE_1, kK	ΔE_2, kK	Ref.
$VO(H_2O)_5^{2+}$	H_2O	16.0	13.0	12
$VO(Cl)_5^{3-}$	$K_3TiCl_5 \cdot 2H_2O$	16.4	15.5	14
$VO(NCS)_5^{3-}$	Nujol mull	17.8	13.7	15
$VO(acac)_2$	Tetrahydrofuran	16.8	15.2	16
$VO(C_2O_4)_2^{2-}$	H_2O	16.6	12.6	17
$VO(acen)$	Tetrahydrofuran	17.7	16.5	18
$VO(salen)$	Tetrahydrofuran	17.0	~16.0	19
$VO(TPP)$	—	—	—	—
$VO(CN)_5^{3-}$	Nujol mull	24.8	15.6	15
$VO(S_2CCN)_2$	—	—	—	—

[a]Abbreviations used: acen—see Fig. 9; salen—bis-(salicylaldehyde)-ethyl-enediimine; TPP—$\alpha,\beta,\gamma,\delta$-tetraphenylporphyrin.

The energy value for TPP^{2-} can be approximated with the value derived from molecular orbital calculations, $\Delta E_1 = 19.5$ kK.[41]

The position of the ΔE_1 maximum should be dependent on the σ-donor strength of the in-plane ligands as well as on the π-donor strength of the same ligands. If the latter quantity does not vary to a large extent, then ΔE_1 should be directly related to the in-plane ligand field strength. Using the values given in Table I with the value given above for TPP the order of the in-plane ligand field strength is given by $CN^- > TPP^{2-} > NCS^- > acen^{2-} > salen^{2-} > acac^- > C_2O_4^{2-} > Cl^- > H_2O$. As expected, the order corresponds quite closely to the spectrochemical series.[32] The position of water is somewhat lower than usually observed. Since we shall see that the in-plane π-donor strength of the ligands does change appreciably for some of the ligands, the magnitude of the σ-interaction must override the former variation. The frequency of ΔE_2 should reflect the difference between the in-plane and axial ligand-to-vanadium π bonding. The order derived from data in Table I is $acen^{2-} > salen^{2-} > CN^- > Cl^- > acac^- > NCS^- > H_2O > C_2O_4^{2-}$. This order does not appear to conform to the π-donating ability of the ligands. The π bonding in transition-metal complexes is usually a relatively minor interaction when compared to the σ-bonding interaction. The amount of axial and in-plane π bonding should also be a function of the σ interaction. This synergistic effect may change the energy levels to the extent that the order given above may not conform exactly to the π-donating ability of the ligand. Data from the ESR spectra shall have a bearing on the point.

CHARACTERISTICS OF THE ESR SPECTRA

It is not our intention to rigorously discuss the theory of ESR of transition-metal complexes, but rather to quote only the useful results.[30] In

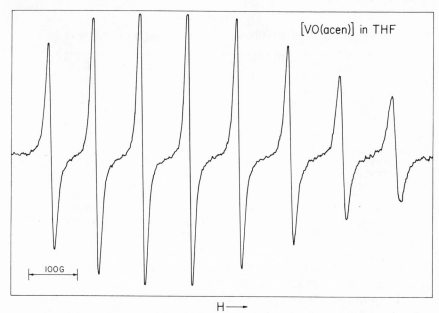

Fig. 3. Electron spin resonance spectrum of bis-acetylacetoneethylenediimineoxovanadium(IV) in tetrahydrofuran at 20°C.

addition to the intensity and line shape of the absorption, two fundamental parameters can be derived from the ESR spectral measurements. These are the Landé effective electron g-factor and the electron–nuclear spin coupling constant A. The isotropic ESR parameters g_0 and A_0 can be determined from the position and spacing of the resonance lines for the room-temperature solution spectrum of the complex. A typical spectrum of an oxovanadium(IV) complex is shown in Fig. 3. In the frozen solid state the axial d^1 case shows two types of resonance components, one set due to the parallel features and the other set due to the perpendicular features. A typical isotropic spectrum is shown in Fig. 4. The isotropic and anisotropic parameters are related by the relations

$$A_0 = (A_{\parallel} + 2A_{\perp})/3 \tag{1}$$

$$g_0 = (g_{\parallel} + 2g_{\perp})/3 \tag{2}$$

The coupling constants are related to the direct dipolar term P (dipole–dipole interaction of electron moment and nuclear moment) and to an indirect dipolar interaction caused by the anisotropy in g values. Neglecting small corrections, the expressions are:

$$A_{\parallel} = -PK - \tfrac{4}{7}\beta_2^{*2}P - (g_e - g_{\parallel})P - \tfrac{3}{7}(g_e - g_{\perp})P \tag{3}$$

$$A_{\perp} = -PK + \tfrac{2}{7}\beta_2^{*2}P - \tfrac{11}{14}(g_e - g_{\perp})P \tag{4}$$

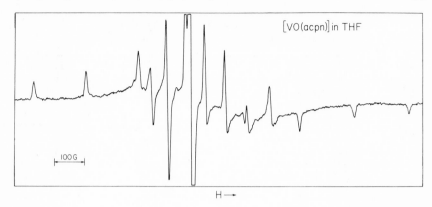

Fig. 4. Electron spin resonance spectrum of bis-acetylacetone(1,2)propylenediimineoxo-vanadium(IV) in tetrahydrofuran at −150°C.

In these equations K is the Fermi contact term, which is related to the amount of unpaired electron density at the vanadium nucleus. Combining Eq. (3) and (4) with Eq. (1) and (2) gives the expression

$$A_0 = -PK - (g_e - g_0)P \tag{5}$$

The g values are generally lower than the free-electron value, $g_e = 2.002$. This lowering can be related to the spin-orbit interaction of the ground state, d_{xy} level, with low-lying excited states. The interaction is given by the following first-order expressions:

$$(g_0 - g_\parallel) = 8(\beta_1^*)^2(\beta_2^*)^2 \xi/\Delta E_1 \tag{6}$$

$$(g_0 - g_\perp) = 2(\beta_2^*)^2(e_\pi^*)^2 \xi/\Delta E_2 \tag{7}$$

where ξ is the spin-orbit coupling constant for the vanadium ion. The other terms in the equations are given in the d-level diagram shown in the previous section.

Isotropic and anisotropic data from ESR spectra of a variety of complexes of oxovanadium(IV) are listed in Table II. The complexes listed were chosen for the reasons stated in the previous section. The spread in g_0 (1.964–1.992) is relatively small in comparison to the variation in A_0 (69.9–118.1 G). In general, the g_0 value-lowering, $g_e - g_0$, gives the inverse of the ΔE_1 ligand order, $H_2O > C_2O_4^{2-} > NCS^- > acac^- > Cl^- > salen^{2-} > acen^{2-} > TPP^{2-} \approx CN^- > S_2CCN^{2-}$. The ligands with the highest ligand field strength give the highest g_0 values. The variation in $g_\parallel$ is somewhat larger than for g_0, i.e., 1.932–1.975. The $g_\parallel$ lowering, $g_e - g_\parallel$, conforms exactly to the g_0 order. The inverse relationship between $g_\parallel$ and ΔE_1 is readily seen by considering Eq. (6). The spread of $g_\perp$ values is the smallest of the three, i.e., 1.976–2.000. If the highest entry is excluded, the range is extremely small (1.976–1.987). As predicted by Eq. (7), $(g_\perp - g_0)$ varies

TABLE II
ESR Parameters for Oxovanadium(IV) Complexes

Complex	Host	$g_0{}^{(a)}$	$g_\parallel{}^{(a)}$	$g_\perp{}^{(b)}$	$A_0{}^{(c)}$, G	$A_\parallel{}^{(c)}$, G	$A_\perp{}^{(d)}$, G	Ref.
$VO(H_2O)_5^{2+}$	$Zn(NH_4)_2(SO_4)_2 \cdot 6H_2O$	1.964	1.932	1.980	118.1	203	78	21
$VO(Cl)_5^{3-}$	$(NH_4)_2InCl_5 \cdot H_2O$	1.971	1.945	1.985	108.9	190	69	22
$VO(NCS)_5^{3-}$	$CHCl_3$	1.967	1.945	1.978	107.0	185	68	23
$VO(acac)_2$	Tetrahydrofuran	1.969	1.945	1.980	106.1	186	67	24
$VO(C_2O_4)_2$	$K_2TiO(C_2O_4)_2 \cdot 2H_2O$	1.964	1.940	1.976	106.0	188	65	17, 40
$VO(acen)$	Tetrahydrofuran	1.974	1.954	1.984	102.2	182	62	18
$VO(salen)$	Tetrahydrofuran	1.973	1.949	1.985	101.7	179	63	19
$VO(TPP)$	$CHCl_3$	1.980	1.966	1.987	97.0	173	59	25
$VO(CN)_5^{3-}$	KBr	1.980	1.972	1.983	83.7	150	51	23
$VO(S_2CCN)_2^{2-}$	$CHCl_3$	1.992	1.975	2.000	69.9	132	43	26

$^a \pm 0.001$.
$^b \pm 0.002$.
$^c \pm 0.5$ G.
$^d \pm 1.0$ G.

approximately in an inverse way to the ΔE_2 ligand order. The ΔE and g ligand orders are not exactly related, since changes in orbital coefficients will also change the g values observed. The isotropic hyperfine coupling constant roughly gives the inverse of the ΔE_1 order, i.e., A_0 is highest for H_2O and lowest for S_2CCN^{2-}: $H_2O > Cl^- > SCN^- > acac^- > C_2O_4^{2-} > acen^{2-} > salen^{2-} > TPP^{2-} > CN^- > S_2CCN^{2-}$. The same order is shown for $A_\perp$ and $A_\parallel$. The spread of $A_\parallel$ is 203–132 G, while it is 78–43 G for $A_\perp$. This represents a change of approximately 30–40%. One reason for the relation of ΔE_1 and the hyperfine splitting constant may arise because of the g_0-value dependence on ΔE_1. Equation (5) predicts that A_0 should depend on g_0. The dependence, however, is slight, since the maximum change in g_0 will change A_0 by only 5 G (observed change, 71 G). This point will be discussed further.

CORRELATIONS

The isotropic ESR values for all the complexes in Table II are plotted in Fig. 5. For convenience of display, the lowering of g_0 from the free-electron value $(g_e - g_0)$ is actually shown. While there is some scatter, it is evident that there is a general trend; as A_0 increases, $(g_e - g_0)$ increases in magnitude. A plot of $g_\parallel$ versus A_0 also shows this trend. On the other hand, a plot of $g_\perp$ against A_0 shows no obvious pattern. From the expression given in the previous section one can readily see that the changes in the g and A values may arise from changes in a number of the molecular parameters and energy-level splittings. The question remains of how these parameters change from complex to complex and what changes dominate the changes in the g and A values.

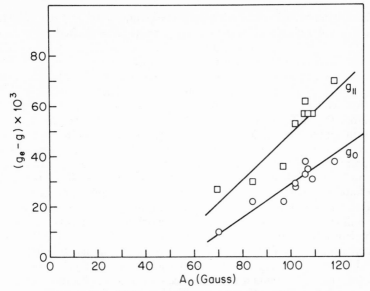

Fig. 5. Plot of ESR parameters for oxovanadium(IV) complexes.

Combining Eq. (3) and (4) and eliminating K, one arrives at an expression to compute values of $(\beta_2^*)^2$ from the experimental results given in Table II. In the computation $A_\parallel$ and $A_\perp$ are taken to be negative in sign, and the value of P is chosen to be $+136$ G.[31] Using Eq. (5) one can also compute values for K. Again A_0 is taken to be negative. Using the g-value expressions and the energy terms from Table I, the other bonding coefficients can be determined. A value of 170 cm^{-1} is assumed for the spin-orbit coupling constant.[24] The bonding coefficient for the d_{z^2} orbital cannot be evaluated, since neither the g nor the A values involve this energy level. The results of the computations are summarized in Table III. The minimum error, which

TABLE III
Isotropic Contact Term and Bonding Coefficients for Oxovanadium Complexes

Complex	K	β_2^{*2}	β_1^{*2}	$(e_\pi^*)^2$
$VO(H_2O)_5^{2+}$	0.83	1.00	0.82	0.84
$VO(Cl)_5^{3-}$	0.77	0.98	0.71	0.79
$VO(NCS)_5^{3-}$	0.75	0.95	0.79	1.01
$VO(acac)_2$	0.75	0.97	0.73	0.82
$VO(C_2O_4)_2^{2-}$	0.74	0.98	0.78	0.98
$VO(acen)$	0.72	0.98	0.64	0.89
$VO(salen)$	0.72	0.94	0.70	0.85
$VO(TPP)$	0.69	0.94	—	—
$VO(CN)_5^{3-}$	0.59	0.82	0.67	1.06
$VO(S_2CCN)_2^{2-}$	0.50	0.73	—	—

arises from experimental uncertainties, can be estimated to be ± 0.01 for each of the entries. A reasonable maximum uncertainty is estimated to be twice the minimum value. Since several of the constants have been chosen rather arbitrarily, the absolute value of the molecular parameters are not significant, only the relative values are of importance. The bonding coefficients listed in Table III are, of course, only approximate values. The neglect of spin-orbit interaction with oxygen and nitrogen ligands is probably justified. On the other hand, the sulfur ligands are extensively covalently bonded and the interaction cannot be neglected. The sulfur ligand bonding coefficients (and CN^-) are then probably too low.[34] The same result arises because of the assumption that ξ is constant. The value should vary to an extent with the amount of covalent bonding. A further source of error is failure to consider additional terms in Eqs. (3)–(7) which are important when appreciable covalent bonding occurs.[30] Finally, misassignment of the electronic spectra will lead to meaningless values of the orbital coefficients.

The isotropic contact term varies considerably from complex to complex, decreasing from a maximum value for the aquo complex to a minimum for the dithiolate. In fact, the K order parallels the A_0 ligand order. The origin of the isotropic contact term has been the subject of some discussion. Since the orbital that contains the unpaired electron, d_{xy}, has zero electron density at the vanadium nucleus and does not mix with the metal $4S$ orbital (in C_{4v} symmetry), there is no direct way of putting unpaired electron density on the nucleus. The nonzero value of K must then arise from an indirect mechanism. McGarvey[31] suggests that the variations can be explained by invoking a spin polarization mechanism. The unpaired electron in the d_{xy} orbital formally creates unpaired electron density in filled $2S$ and $3S$ orbitals of the vanadium. In the absence of covalent bonding and $4S$ mixing, spin polarization should remain constant for all vanadium complexes and be equal to the free-ion value, K_0. Taking into account covalent bonding, K should depend on the d-orbital population for the unpaired electron:

$$K \approx (\beta_2^*)^2 K_0 \tag{8}$$

A plot of K vs $(\beta_2^*)^2$ is shown in Fig. 6. The plot does not appear to give the predicted linear correlation. Actually, a line can be drawn from the origin through the points representing $VO(S_2CCN)_2^{2-}$ and $VO(CN)_5^{3-}$. These complexes are expected to be the most covalent of all those shown. The other points would then be to the right of this line. This deviation might be due to a second contribution to the contact term. It must be admitted that variation in $(\beta_2^*)^2$ is small for most of the complexes (0.94–1.00), and it is only large for the S_2CCN^{2-} and CN^- complexes. This is consistent with Kivelson's conclusion that the d_{xy} orbital is essentially nonbonding and that $(\beta_2^*)^2$ remains approximately constant for most complexes. Lowering of the $(\beta_2^*)^2$ value arises from delocalization of the electron onto the ligand with the increase in covalent bonding. The delocalization occurs via in-plane π-bonding of the d_{xy} orbital with the π-orbitals of the basal ligands. While there probably is some π-interaction for all ligands, the effect only becomes

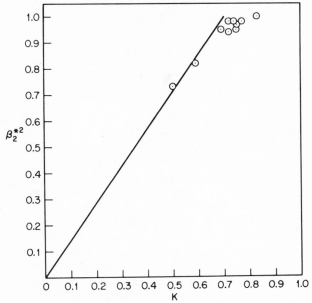

Fig. 6. Plot of isotropic contact term vs the d_{xy} coefficient for oxovanadium(IV) complexes.

appreciable with good π-bonding ligands. In fact, the values of $(\beta_2^*)^2$ follow the general π-bonding order (nephelauxetic) of ligand donor atoms,[32] $S > C > N > Cl \geqslant O$.

The value of the in-plane σ-bonding coefficients $(\beta_1^*)^2$ generally follows the σ-donor strength of the ligand, i.e., $(\beta_1^*)^2$ decreases as the covalent bonding increases, $C \geqslant N > Cl > O$. The variation of $(e_\pi^*)^2$ with ligand is quite surprising. It appears that the most delocalization occurs in the chloro complex, i.e., the complex with the weakest in-plane donors (both π and σ). In general, the stronger is the in-plane donor atom, the less covalent is the axial vanadium–oxygen bond. This can be rationalized by considering charge build-up on the metal. As the charge on the metal increases, the charge donation from the axial oxygen decreases (π-bond strength decreases). The infrared vibrational frequency of V=O follows this order, being highest for those complexes with the weakest in-plane donor atoms.[33] The anisotropy in g, $\Delta g = g_\perp - g_\parallel$, should be a function of the difference in the axial and in-plane ligand fields. If the charge build-up concept is correct, then as the in-plane ligand field increases, the axial ligand field should decrease. The Δg should therefore be smaller for strong-field ligands binding to the in-plane positions of oxovanadium(IV). The relevant experimental evidence to compare Δg with would be $\Delta E_3 - \Delta E_1$ from the electronic spectra. Unfortunately, ΔE_3 is not known with any certainty. All other things being equal, ΔE_1 should give the in-plane ligand field. As predicted, Δg does correlate with ΔE_1, i.e., as Δg decreases, ΔE_1 increases.

In fact, the Δg order for the ligands is exactly the same as the ΔE_1 order given in the previous section.

The difference between the dependence of the isotropic contact term on $(\beta_2^*)^2$ as predicted by Eq. (8) and as observed is not large. The increase represents a maximum difference of approximately 15%. The small increase in the value of K can arise from participation of the empty $4S$ orbital on the metal in σ-bonding to the ligands. Both the empty $4S$ and $d_{x^2-y^2}$ orbitals could effectively overlap with the filled σ levels of the basal ligands. The MO formed with $4S$ orbitals should put partial $4S$ density in a filled bonding orbital. This in turn should undergo spin polarization by the d_{xy} electron. There is no straightforward way of assessing the contribution of this effect. One way of getting around this difficulty is to consider the σ bonding of the metal to ligand for the d orbitals. For the C_{4v} symmetry case the $4S$ and $3d_{x^2-y^2}$ orbitals do not mix, while the $4S$ and d_{z^2} do mix. On the other hand, for the lower symmetry, C_{2v}, both the d levels mix with the $4S$ orbitals. This would, of course, introduce $4S$ character into the σ-bonding system of the complex for both the axial and in-plane σ ligand-to-metal bonds. Whether this mixing occurs is not of great importance, since the $4S$ σ bonding should parallel the d σ bonding. The extent of the ligand-to-metal interaction should be related to both the energy of the antibonding $d_{x^2-y^2}$ level and its orbital coefficient. The $4S$ contribution to K should be proportional to the metal electron density in the filled orbital that contains the contribution from the $4S$ orbital. The delocalization in the σ system of the complex is expressed by the bonding coefficient for the $d_{x^2-y^2}$ level. It is assumed that the $4S$ coefficient

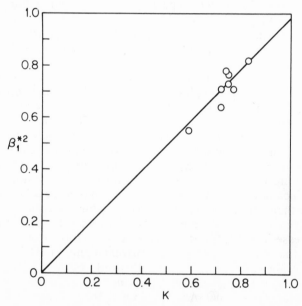

Fig. 7. Plot of isotropic contact term vs the $d_{x^2-y^2}$ orbital coefficient for oxovanadium(IV) complexes.

would follow the $d_{x^2-y^2}$ coefficient. A plot of the isotropic contact term vs $(\beta_1^*)^2$ is shown in Fig. 7. Although there is some scatter, it is apparent that the points fall near a straight line that passes through the origin. The $4S$ contribution to the isotropic contact term appears to be important only for the less-covalently-bound ligand donors. There would be, however, some $4S$ contribution to K for all ligand types.

The variation in K is in the range 0.83–0.72 for all the ligands except the last three in Table III. It has been shown that these latter ligands are appreciably involved in covalent bonding to the metal. Since the change in $(\beta_2^*)^2$ is small, the major change in K for these less-covalently-bound complexes must be due to the $4S$ σ-bonding effect. The energy separation of the bonding and antibonding orbitals for the $4S$ ligand interaction is inversely proportional to the indirect $4S$ contribution to K.[34] This energy separation should be a function of the in-plane ligand field. The electronic absorption spectra give an equivalent piece of information, ΔE_1. In other words, as the in-plane ligand field increases, as measured by ΔE_1, the $4S$ bonding contribution to the isotropic contact term decreases. A plot of $1/\Delta E_1$ vs K (Fig. 8) does indeed show the proper relationship. The value used for VO(TPP) was the calculated one,[41] while the value for VO(S$_2$CCN)$_2^{2-}$ was the observed weak absorption maximum in the near UV. A direct contribution to K can arise from d_{xy}, $4S$ mixing in low-symmetry complexes (less than C_{2v}). This would be of opposite sign to the indirect $4S$ contribution. The relatively small values of A_0 for low-symmetry complexes of oxygen and nitrogen donors probably arises from this effect.[40]

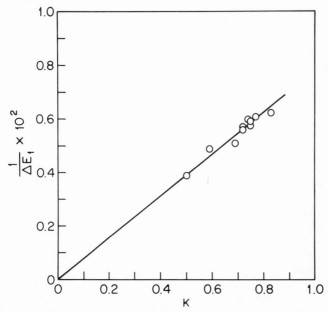

Fig. 8. Plot of isotropic contact term vs the inverse of ΔE_1, the electronic absorption maxima.

Several ESR correlations have previously been suggested for transition-metal complexes. For example, the isotropic contact term increases as the electronegativity difference between the ligand atom and the metal increases.[42] This has been interpreted to mean that the Fermi contact term decreases as the covalency of the metal–ligand linkage increases. The dependence of K on the donor atom also seems to be related to the electronegativity effect for the oxovanadium complexes. For instance, the isotropic contact term decreases as the electronegativity of the donor atom decreases (S < C < N < Cl < O). The same dependence of K on covalency is, of course, given in a much more straightforward way by Figs. 6 and 7. A crystal field interpretation has been given to explain the A_0 variation for oxovanadium(IV) complexes.[29] There appears to be a linear correlation between the isotropic coupling constant and the ratio of axial to equatorial charge (as obtained from electronic spectra). An increase in A_0 is then related to a decrease in the ratio of axial to equatorial charge. All other things remaining equal, this just states that as the in-plane ligand field increases, A_0 decreases. This is equivalent to the relationship shown in Fig. 8.

One can now readily see the reason for the dependence of A_0 on $g_{\parallel}$ as shown in Fig. 5. The Fermi contact term is directly proportional to the A_0 (with a small correction). We have previously shown that in a rough way K is directly proportional to $(\beta_2^*)^2$ and $(\beta_1^*)^2$ and also that K is inversely proportional to ΔE_1. From Eq. (6) we see that these are just the quantities that make up $g_{\parallel}$. The $g_{\perp}$ term does not vary to a great extent from complex to complex. Further, this quantity is related mostly to the axial V=O interaction and is not directly related to the in-plane ligands. Since g_0 parallels the changes in $g_{\parallel}$, then one can explain the general trend of g_0 vs A_0 as seen in Fig. 4. In addition to this ligand field effect, the trend is related qualitatively to the amount of electron delocalization (covalent bonding) occurring in the complex. If the unpaired electron is not confined to the vanadium atom, spin-orbit coupling is reduced and g_0 values increase. Furthermore, interaction with the metal nucleus is reduced, and this leads to a smaller hyperfine splitting.

SUBSTITUENT EFFECTS

In general, the ESR spectra of VO^{2+} complexes are strongly dependent on the donor atoms bound to the metal; e.g., they are different for nitrogen and oxygen donors. There are also observable, but smaller, differences with different types of ligands that contain the same donor atom. For example, β-diketone ligands are more covalently bound than water ligands, and consequently the g values are higher and hyperfine splittings lower. However, substituent changes within one class of a particular donor-atom ligand give rise to a small variation in g and A values.

Table IV lists data from the ESR spectra of some oxovanadium complexes of β-ketimines. Figure 9 gives the structural representations and abbreviations used here. It is first noted that the g values remain invariant

TABLE IV
ESR Parameters for Oxovanadium(IV) Complexes of β-Ketimines in Tetrahydrofuran[18,19]

Complex	$g_0{}^{(a)}$	$g_\perp{}^{(b)}$	$g_\parallel{}^{(a)}$	$A_0{}^{(c)}$, G	$A_\parallel{}^{(c)}$, G	$A_\perp{}^{(d)}$, G
VO(acen)	1.974	1.984	1.954	102.2	182	62
VO(acpn)	1.974	1.984	1.954	102.6	183	62
VO(bzen)	1.975	1.986	1.952	102.6	183	62
VO(bzpn)	1.975	1.984	1.957	103.4	182	64
VO(tfen)	1.974	1.983	1.955	104.6	183	63

$^a \pm 0.001.$
$^b \pm 0.002.$
$^c \pm 0.5$ G.
$^d \pm 1.0$ G.

to substitution, even for the strongly-electron-withdrawing trifluoromethyl group. In general, this has been found for other ligand types in oxovanadium complexes. For example, fluorine or phenyl substitution does not affect the g values of the β-diketone complexes.[29] Further, for the porphyrin-type complexes, mesoporphyrin IX, protoporphyrin IX, etioporphyrin II, and α,β,γ,δ-tetraphenylporphyrin have the same g values.[19] Even a rather

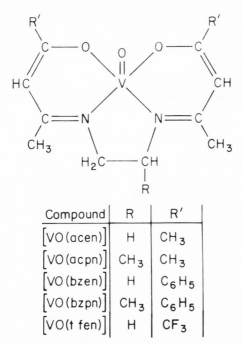

Compound	R	R'
$[VO(acen)]$	H	CH_3
$[VO(acpn)]$	CH_3	CH_3
$[VO(bzen)]$	H	C_6H_5
$[VO(bzpn)]$	CH_3	C_6H_5
$[VO(tfen)]$	H	CF_3

Fig. 9. Structural representation for β-ketimine complexes of oxovanadium(IV).

drastic substituent change, as occurs in going to the phthalocyanines, does not observably change the g values.[35] It is evident from the small changes in absorption spectra with substituent that the g values should change. Unfortunately, the magnitude of the change is within the experimental uncertainty of most measurements. The hyperfine splittings are sensitive to substituent effects. The isotropic coupling constant changes by as much as approximately 2.5 G with trifluoro methyl substitution. For β-diketones the change can be somewhat larger, approximately 4.5 G.[29] For the porphyrin-type ligands the changes are smaller. The changes in the hyperfine splitting constant are equivalent to a change in the isotropic contact term of from 0.01 to 0.03. Since covalent bonding should be about equal in these complexes, the change is due to the indirect $4S$ contribution to K. The increase in A_0 of approximately 4.5 G for the hexafluoroacetylacetone complex of VO^{2+} implies that electron-withdrawing fluorine substituent causes the in-plane σ and π bonding to decrease. This, of course, correlates well with the decreased donor strength of fluorinated ligands.[36] Except for the most-powerful electron-withdrawing or electron-donating groups, the changes in A_0 are usually small.

SOLVENT EFFECTS

Moderately large solvent effects are noted for the ESR spectra of oxovanadium(IV) complexes. In fact, the magnitude of the change is as great or greater than that of the substituent effects. Data for VO(acac)$_2$ in a number of solvents are summarized in Table V. The spectral data are

TABLE V
ESR Parameters and Visible Absorption Maxima for VO(acac)$_2$ in Various Solvents

Solvent[a]	g_0 [b]	A_0 [c], G	ΔE_1 [d], kK	ΔE_2 [d], kK	Ref.
CS$_2$	1.968	108.3	16.7	15.1	24
Benzene	1.970	108.2	16.9	15.3	28
Toluene	1.969	107.0	16.9	15.3	27
CHCl$_3$	1.970	106.4	16.9	14.9	24
THF	1.969	106.1	16.8	13.7	24
Acetone	1.968	106.0	16.7	14.3	16
CH$_3$CN	1.969	105.3	16.8	14.2	16
Nitrobenzene	1.972	104.6	16.7	14.7	16
DMF	1.968	104.4	17.0	13.0	29
Pyridine	1.970	104.4	17.4	13.0	28
CH$_3$NH$_2$	1.969	103.9	17.4	13.1	24
NH$_3$	1.968	103.4	—	—	24
Methanol	1.969	102.3	17.4	13.0	28

[a]THF: tetrahydrofuran; DMF: dimethylformamide.
[b]± 0.001.
[c]± 0.5 G.
[d]Selbin.[20]

typical of the solvent effects noted for VO^{2+} complexes. Within experimental error the g_0 value does not vary. On the other hand, the A_0 values vary over a wide range ($\Delta A_0 \sim 5.0$ G). The highest values of A_0 are found for the complexes in noncoordinating solvents. In midrange the poorly-coordinating solvents are found. The lowest values of A_0 are found for good donating solvents (basic). Another factor is the hydrogen-bonding ability of the solvent molecule. For example, the largest shift is seen for methanol, even though it is reported to be a thermodynamically weaker donor than pyridine. Further, the hydrogen-bonding effect undoubtedly gives rise to the relatively large ΔA_0 observed for $CHCl_3$, CH_3NH_2, and NH_3. Another factor is the dielectric constant of the medium. It appears that ΔA_0 increases as the dielectric constant of the medium increases.[28] This effect is probably related to the dipolar nature of the complex.

The geometry of the oxovanadium complexes is such that a sixth ligand molecule (solvent) can readily be added to the vanadium along the z axis *trans* to the axial oxygen. Although weak, the interaction enthalpy is still substantial (-5 to -10 kcal/mole).[37] Good donor solvents should then add to the vanadium and increase the ligand field along the z axis. This interaction should weaken the V=O bond. In agreement with this, infrared spectral measurements have shown that $v(V{=}O)$ is lowered by the interaction.[33] Further, ΔE_2 is generally lowered in energy by the decrease in V=O π bonding. Weakening of the axial bond allows a stronger in-plane σ and π interaction. It is this change which gives rise to the lowering of A_0. The increase in the in-plane donation is also evidenced by the increase in the energy of the ΔE_1 band. In fact, the K values for all the entries in Table V fall on the plot given in Fig. 8.

Hydrogen-bonding effects can likewise accomplish the same result. Binding of the solvent molecule to the axial oxygen atom decreases the oxygen-to-vanadium donation along the z-axis and gives rise to a greater in-plane donation. The hyperfine splitting then decreases in magnitude. Again ΔE_1 increases in energy and ΔE_2 decreases in energy as a result of the axial-oxygen hydrogen-bonding interaction.

Solvent effects appear to be somewhat smaller for β-ketimine and porphyrin-type ligands than for the β-diketones. This can be rationalized by taking into account that there is already a strong in-plane interaction and weaker axial bonding in these complexes.

APPLICATION TO VANADIUM IN PETROLEUM

The nature of the nonporphyrin vanadium in asphaltene still remains a mystery. Knowledge about the chemical nature of the metal atom is a prerequisite to understanding not only the role that the metal ion plays in the origin of petroleum, but also to determine how to remove this deleterious element from petroleum products. It has been suggested that nonporphyrin vanadium is bound to hetero atoms such as nitrogen, oxygen, and sulfur in the polyaromatic portion of asphaltene.[38] By observing the ESR spectra of

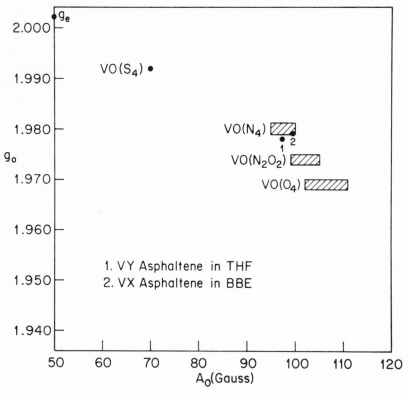

Fig. 10. Plot of ESR parameters for model complexes of oxovanadium(IV).

the vanadium chelates *in situ* or in fractions of asphaltene, one should be able to learn something about how the metal is bound in the macromolecules. The g_0 and A_0 values determined experimentally can then be used with the correlation found here to partially answer this question. The ESR data for aromatic nitrogen, oxygen, and sulfur donors are displayed in Fig. 10. The model compounds used to construct the plot are as follows: $VO(O)_4$ represents the β-diketones; $VO(N_2O_2)$ represents the β-ketimines and salicylaldimines; $VO(N)_4$ represents the tetraphenylporphyrins, porphyrins, and the phthalocyanines; and finally $VO(S)_4$ represents the data for $VO(S_2CCN)_2^{2-}$. The shaded boxes give the spread of experimental values for the different classes of complexes in different solvents. As was pointed out before, the variations in g_0 are small and the plot actually represents the experimental uncertainty in the measurements. Conversely, A_0 changes substantially with solvent and substituent. The ESR parameters for the $VO(N_3O)$ and $VO(NO_3)$ can be predicted from the values for the model compounds. The $VO(N_3O)$ ligand type should be at $g_0 \approx 1.977$ and $A_0 \approx 100\,G$, midway between $VO(N_4)_4$ and $VO(N_2O_2)$. In the same way the $VO(NO_3)$ ligand type should fall at $g_0 \approx 1.972$ and $A_0 \approx 104\,G$. As was pointed out earlier, only the g_0

values are of reasonable predictive value, since they are dependent on ligand type and are invariant to substituent and solvent effects. The isotropic coupling constants can, however, vary as much as ± 2 G under similar conditions. Consequently, low values for A_0 for a particular value of g_0 would be indicative of axial coordination or hydrogen bonding of the solvent to the VO^{2+} ion. On the other hand, high values of A_0 would be indicative of the presence of electron-withdrawing groups near the donor atoms. Donor sites with one or more sulfur atoms, with the remaining being nitrogen donors, would clearly fall in the ranges $g_0 = 1.992–1.980$ and $A_0 = 70–97$ G. For example, $VO(N_2S_2)$ would fall at $g_0 = 1.986$, $A_0 = 83$ G. Therefore donor sites containing only sulfur and nitrogen should clearly be distinguishable from those containing nitrogen and oxygen, etc. This discussion would, of course, rest on a better foundation if data for the $VO(N_2S_2)$ model compound system were available. Similar reasoning can be applied to the oxygen–sulfur mixed-ligand complexes. The above discussion is based on the well-known rule of average environment in ligand field theory.[31]

The graph in Fig. 10 shows two points for the experimentally determined ESR spectra of vanadium in asphaltene. It is seen that these points fall within and close to the $VO(N)_4$ case. This shows that, at least in this case, only the porphyrin vanadium is being observed. Any other lines are much weaker and obscured by the broad lines of the main species. An alternate conclusion is that the nonporphyrin vanadium is bound in sites that are very much like a porphyrin ligand (four nitrogen donors). This appears to be ruled out on the basis of a statistical mass-spectral study of asphaltic resins.[43] These data show that the major nitrogen-containing materials have odd nitrogen numbers, perhaps N_3 with either sulfur or oxygen. The ESR spectrum of solid asphaltene shows a number of lines in addition to the ones that correspond to the porphyrin complexes. These lines are weaker than the main anisotropic features and are most easily noted toward the center of the spectra.[39] Although there are other good possibilities, one can say that these lines may be due to vanadium complexes bound in environments different from the porphyrin. After extracting the porphyrin vanadium it would be interesting to fractionate the asphaltene, concentrate the vanadium, and then observe the ESR spectra. In this way the ESR spectra of the nonporphyrin vanadium might be observable.

It is hoped that the correlations established in this chapter will be of some use in the investigation of vanadium in this and other complex systems.

ACKNOWLEDGMENT

The authors thank Mr. Timothy R. Drury, Research Services, Mellon Institute, for his assistance in some of the experimental work.

This work was sponsored by Gulf Research & Development Company as part of the research program of the Multiple Fellowship on Petroleum.

REFERENCES

1. A. Carrington and A. D. McLachlan, *Introduction to Magnetic Resonance* (Harper and Row, New York, 1967).
2. J. Peisach and W. E. Blumberg, this volume, Chapter 6.
3. H. J. M. Bowen, *Trace Elements in Biochemistry* (Academic Press, New York, 1966).
4. H. J. Bielig and E. Bayer, *Ann. Chem.* **580**, 135 (1953).
5. D. Bertrand, *Bull. Am. Mus. Nat. Hist.* **94**, 409 (1950).
6. O. A. Radchenko and L. S. Sheshiha, *Visn. Neft. Nauch.-Issled. Geol. Inst.* **33**(1), 274 (1955).
7. A. J. Saraceno, D. T. Fanale, and N. D. Coggeshall, *Anal. Chem.* **33**, 500 (1961).
8. E. W. Baker, T. F. Yen, J. P. Dickie, R. E. Rhodes, and L. F. Clark, *J. Am. Chem. Soc.* **89**, 3631 (1967).
9. J. Selbin, *Coord. Chem. Rev.* **1**, 293 (1966).
10. R. P. Dodge, P. H. Templeton, and A. Zalkin, *J. Chem. Phys.* **35**, 55 (1961).
11. P. Kierkegaard and J. M. Longo, *Acta Chem. Scand.* **19**, 1906 (1965).
12. C. J. Ballhausen and H. B. Gray, *Inorg. Chem.* **1**, 111 (1962).
13. K. M. Jones and E. Larson, *Acta Chem. Scand.* **19**, 1210 (1965).
14. R. A. D. Wentworth and T. S. Piper, *J. Chem. Phys.* **41**, 3884 (1964).
15. J. R. Wasson, *J. Inorg. Nucl. Chem.* **20**, 171 (1968).
16. J. Bernal and P. H. Rieger, *Inorg. Chem.* **2**, 256 (1963).
17. R. M. Golding, *Mol. Phys.* **5**, 369 (1962).
18. L. J. Boucher, E. C. Tynan, and T. F. Yen, *Inorg. Chem.* **7**, 731 (1968).
19. L. J. Boucher, E. C. Tynan, and T. F. Yen, *Inorg. Chem.* **8** (1969).
20. J. Selbin, *Chem. Rev.* **65**, 153 (1965).
21. R. H. Borcherts and C. Kikuchi, *J. Chem. Phys.* **40**, 2270 (1964).
22. K. DeArmond, B. B. Garrett, and H. S. Gutowsky, *J. Chem. Phys.* **12**, 1019 (1965).
23. H. A. Kuska, Thesis, Michigan State University, 1965.
24. D. Kivelson and S. K. Lee, *J. Chem. Phys.* **41**, 1896 (1964).
25. J. M. Assour, *J. Chem. Phys.* **43**, 2477 (1965).
26. N. M. Atherton, J. Locke, and J. A. McCleverty, *Chem. and Ind.* **1965**, 1300.
27. R. Wilson and D. Kivelson, *J. Chem. Phys.* **44**, 154 (1966).
28. F. A. Walker, R. L. Carlin, and P. H. Rieger, *J. Chem. Phys.* **45**, 4181 (1966).
29. H. A. Kuska and M. T. Rogers, *Inorg. Chem.* **5**, 3113 (1966).
30. B. R. McGarvey, "Electron Spin Resonance of Transition Metal Complexes," in: *Transition Metal Chemistry*, Vol. 3, R. L. Carlin, ed. (Marcel Dekker, New York, 1966).
31. B. R. McGarvey, *J. Phys. Chem.* **71**, 51 (1967).
32. G. K. Jorgensen, *Absorption Spectra and Chemical Bonding in Complexes* (Pergamon Press, Oxford, 1962).
33. J. Selbin, L. H. Holmes, and S. P. McGlynn, *J. Inorg. Nucl. Chem.* **25**, 1359 (1963).
34. B. R. McGarvey, *J. Chem. Phys.* **41**, 3743 (1964).
35. J. M. Assour, J. Goldmacher, and S. E. Harrison, *J. Chem. Phys.* **43**, 159 (1965).
36. A. E. Martell and M. Calvin, *Chemistry of the Metal Chelate Compounds* (Prentice Hall, Englewood Cliffs, New Jersey, 1952), p. 549.
37. R. L. Carlin and F. A. Walker, *J. Am. Chem. Soc.* **87**, 2128 (1965).
38. T. F. Yen, J. G. Erdman, and A. J. Saraceno, *Anal. Chem.* **34**, 694 (1962).
39. T. F. Yen, E. C. Tynan, G. B. Vaughan, and L. J. Boucher, in: *Advances in Spectroscopy of Fuels*, R. A. Friedel, ed. (Plenum Press, New York, 1969).
40. K. Wuthrich, *Helv. Chim. Acta* **48**, 779, 1012 (1965).
41. M. Zerner and M. Gouterman, *Inorg. Chem.* **5**, 1699 (1966).
42. R. S. Titte, *Phys. Rev.* **131**, 623 (1963).
43. J. P. Dickie and Teh Fu Yen, *American Chemical Society, Div. Petroleum Chemistry, Preprints* **13**(2), F140 (1968).

Computer-Assisted Analysis of the EPR Spectrum of $Cu^{2+}:ZnF_2$*

H. M. Gladney, B. Johnson, and J. D. Swalen

IBM Research Laboratory
Monterey & Cottle Roads, San Jose, California

The EPR spectrum of a Cu^{2+}-doped single crystal of zinc fluoride has been measured at X band. This was done by utilizing a laboratory automation system for the IBM 1800 by which we are able to control the experimental observations and obtain improved results. Other computer programs were applied to analyze the spectra for the magnetic parameters, i.e., the g values, the copper hyperfine structure constants, and the fluorine superhyperfine structure constants. Covalency measures were then derived from these values. The details of the spectrum and its analysis, the laboratory automation system, and the covalency parameters compared with other copper compounds will be discussed.

Although the properties of copper compounds, both in the doped inorganic crystals and in organic complexes, have been studied extensively by electron paramagnetic resonance and optical spectroscopy (see, e.g., McGarvey[1] or Orton[2]), the information obtained in any given case has usually been insufficient to ascertain with any reliability the chemical bonding parameters. We undertook the study of a single crystal of copper doped in zinc fluoride† with the expectation that the superhyperfine structure from the fluoride ions would make possible a more complete analysis. Zinc fluoride (as is cupric fluoride) has been determined[3,4] to be of the rutile structure, i.e., two zinc ions per unit cell, oriented at 90° from one another, and each surrounded by six fluoride ions. Our experimental results show that the impurity copper ions are substitutional for zinc ions. Since the two axial fluoride ions are closer than the equatorial ones, crystal field theory suggests that the ground state should be the d_{z^2} orbital. Also, since the equatorial ions do not form a square, but are compressed such that one F—Zn—F angle is approximately 40° and the other is approximately 50°, rhombic anisotropy of the resonance spectra is expected. The EPR spectrum was recorded at 1.2°K and 77°K with a conventional X-band, 400-Hz modulation superheterodyne spectrometer[5] at various crystal angles with respect to the magnetic field.

*More detailed accounts of both the EPR spectrum and the laboratory system will be published elsewhere.
†We wish to thank R. S. Feigelson of Stanford University for supplying us with this crystal.

The spectrum was observed with the assistance of an IBM 1800 computer running under a time-sharing monitor system.[6] Digital output from the computer controlled the magnetic field, the orientation of the 12-in. rotating electromagnet, and the gain, time constant, and modulation amplitude of the phase-sensitive detector. The output voltage from the detector was transmitted on a twisted-pair of conductors to the computer some 300 ft away. Analog-to-digital conversion was accomplished there. A sequence of tasks, such as recording, plotting, digital smoothing, baseline drift correction, time averaging, line sharpening, and magnet rotation, were specified at the initiation of each set of observations, with options for sequence modification by the experimenter during execution. Data points were read and recorded when a computer timer expired, subject to the arithmetic unit of the computer being free from higher priority tasks. Since a scanning spectrometer is inherently a low-data-rate device, and since time is not the independent variable, the unpredictable delays inherent in this method are irrelevant to the experiment. To the experimenter some salient features of our "on-line" experiment are:

1. No noticeable delay occurs between local and computer operation.

2. A long experiment may be set up for unattended operation, so that the tedium of searching under high sensitivity for extended periods of time is removed.

3. The data are plotted accurately (0.01 in.) and on the same scale as theoretical calculations. Hence, precise comparison for parameter fitting is very easy.

4. The record of the experiment is typed on an on-line typewriter located in the laboratory for direct insertion into a notebook.

5. A local and direct plot is available for immediate checking and supervision by the experimenter.   .

Our initial spectral measurements indicated the expected two substitutional sites for copper ions. With the magnetic field along a (110) direction in the lattice we saw the resonance from the axial direction (z) and one equatorial direction (x). In the (001) direction both sites are equivalent and one spectrum (y) was observed. In the axial direction (z) hyperfine and fluoride superhyperfine structure were clearly seen and the g value was close to 2 as expected for d_{z^2} ground state. The equatorial spectra exhibited only one major structural feature, but fluoride hfs was present as evidenced by the serrated line shapes. The g values were shifted from 2 by orbital-momentum contributions and were, although not identical, quite close to one another.

Computer programs, described elsewhere,[7] were used to calculate trial spectra, estimate the magnetic parameters, and plot the results for comparison with the experimental results. Surprisingly, the superhyperfine structure of the axial fluoride ions dominated the copper hyperfine structure. Because we sought a detailed comparison of the total line shape of each pattern (a necessary condition to ensure accuracy of the magnetic parameters) over 100 trial computer-plotted spectra were made. Our determined "best values" for the magnetic parameters at 1.2°K are as shown in Table I.

TABLE I

	X	Y	Z
g	2.4275	2.354	2.061
A_{Cu}	18×10^{-4} cm^{-1}	8	49.9 ± 0.5 (^{63}Cu)
			53.5 ± 0.5 (^{65}Cu)
$A_{axial F}$	55	48	121.8
$A_{equatorial F}$	38	42	13.

From the constants $|\psi_{2s}(0)|^2 \approx 70 \times 10^{24}$ cm^3 and $\langle r^{-3} \rangle_{2p} \approx 44 \times 10^{24}$ cm^3 for Cu^{2+}, the values for the s,σ, and dipole–dipole contributions to the superhyperfine terms for electron transfer are $A_s^0 \approx 1.44$ cm^{-1}, $A_\sigma^0 \approx 0.044$ cm^{-1}, and $A_{dd}^0 \approx 0.00016$ cm^{-1}. After subtracting the A_{dd}^0 term, the experimental values in the axial direction are $A_s \approx 77.2 \times 10^{-4}$ cm^{-1} and $A_\sigma \approx 208 \times 10^{-4}$ cm^{-1}. Fractional values are therefore $f_s \approx 1\%$ and $f_\sigma \approx 6\%$, both being somewhat larger than previously reported values for other transition-metal ions.[8]

REFERENCES

1. B. R. McGarvey, "Electron Spin Resonance of Transition-Metal Complexes," in: *Transition Metal Chemistry*, Vol. 3 (Marcel Dekker, New York, 1966), p. 89–201.
2. J. W. Orton, *Rept. Prog. Phys.*, **22**, 216 (1959).
3. W. H. Baur, *Acta Cryst.* **11**, 488 (1958).
4. C. Billy and H. M. Haendler, *J. Am. Chem. Soc.* **79**, 1049 (1957).
5. J. P. Goldsborough and T. R. Koehler, *Phys. Rev.* **133**, A135 (1964).
6. H. M. Gladney, B. Johnson, and T. Kuga, "Computer-Assisted Spectroscopy," *IBM J. Res. Dev.* **13**, 36 (1969).
7. J. D. Swalen and H. M. Gladney, *IBM J. Res. Dev.* **8**, 515 (1964); H. M. Gladney, *Phys. Rev.* **143**, 198 (1966).
8. J. Owen and H. J. M. Thornley, *Rept. Prog. Phys.* **29**, 675 (1966).

Second-Derivative Line-Sharpening Device for Electron Paramagnetic Resonance

David E. Wood

Department of Chemistry and Mellon Institute
Carnegie-Mellon University
Pittsburgh, Pennsylvania

A second-derivative line-sharpening device is described which is capable of reducing the apparent linewidth of an EPR signal by half. The principle of operation is field modulation at odd subharmonics of the second field modulation frequency. Accumulation of the sharpening terms in a CAT is found to greatly improve the signal-to-noise ratio of the sharpened spectra and to reduce the set-up time required.

INTRODUCTION

Electron paramagnetic resonance (EPR) spectra usually consist of many overlapping lines, and the EPR spectroscopist must establish the number, position, and relative intensity of these lines in order to interpret the spectrum. Resolution enhancement is very often desirable in the analysis of these overlapping spectral lines. Processes for resolution enhancement can be generally divided into two classes: (1) those which change the basic shape of the spectrum, and (2) those which do not. An example of the former type of method is the taking of multiple derivatives. The latter is usually called *spectrum sharpening*. Analog differentiation by operational amplifier circuits[1] has been used to take multiple derivatives of EPR spectra. This technique, while allowing better resolution of line positions, increases the number of zero crossings to the point that complex spectra can get very difficult to decipher. The analog method is also rate dependent, which makes it difficult to utilize when different linewidths occur in the same spectrum. Digital methods of taking multiple derivatives (see, e.g., Hedberg and Ehrenberg[2]) suffer from a time lapse between taking the spectra and seeing it in its treated form. Both of these methods of resolution enhancement require careful filtering to ensure that noise spikes are not treated as signal. Spectrum sharpening which has been done by filtering the signal[3] to introduce sharpening terms is also scan-rate dependent. All of the methods mentioned above are essentially treatments of the signal data after the signal has been received, in that these methods could just as well be used on prerecorded data. The method of phase detection at harmonics of the field modulation frequency, or, alternately, modulation at subharmonics of the phase-sensitive detection frequency, must, however, be done during the experiment, and operates directly on the line shape of the absorbing sample. The instrumental technique of subharmonic modulation for adding sharpening terms to an EPR spectrum was successfully applied by Glarum[4] to the first-derivative presentation. This method is scan-rate independent and requires no special filters in the detection circuit in order to eliminate sharpening of the noise spikes as in the other methods. There is, however, a loss in signal-to-noise ratio of between 75 and 100 for each sharpening term added. This loss can be prohibitive in many cases.

The method of resolution enhancement described in this chapter is a combination of the two classes previously mentioned. The second-derivative mode of presentation was chosen as a basis for spectrum sharpening for the following reasons: (1) Baseline drift is greatly reduced, which improves the result of sharpening; (2) taking the second derivative enhances the resolution in itself, with only the loss of about 2 in signal-to-noise ratio; and (3) incompletely-resolved peaks are easier to identify and measure than baseline crossings would be under the same circumstances. The second-derivative presentation is obtained in the usual way[5] with the first phase-sensitive detector (Varian V-4560) operating at 100 kHz with 2-kHz bandwidth and the second phase-sensitive detector (PAR HR-8) operating at 750 Hz. The sharpening terms are obtained by field modulation at odd subharmonics of the 750-Hz frequency. The sharpening terms can be added simultaneously with the second-derivative signal to yield a sharpened second-derivative representation on the chart recorder in the usual manner. This is a satisfactory method if the signal is strong and not many sharpening terms are desired; however, it was found that combination of the second derivative and the individual sharpening terms could best be effected by adding them sequentially in the memory of a CAT by means of successive scans. In this manner the progress of the spectrum sharpening can be observed with the addition of each term; simultaneous adjustment of all subharmonic phases and amplitudes is not necessary, and, best of all, the loss of signal-to-noise ratio for each sharpening term added is small compared to the simultaneous addition method.

EXPERIMENTAL SETUP FOR SECOND-DERIVATIVE LINE SHARPENING

The experimental setup is illustrated schematically in Fig. 1. For direct presentation of the sharpened EPR signal the sample undergoes simultaneous field modulation at 100 kHz, 750 Hz, 250 Hz, 150 Hz, and 107 Hz. The PSD reference phases for the 100 kHz and 750 Hz are adjusted in the usual manner and the subharmonic phases are adjusted in the modulation channels. Considering only the first two modulation frequencies for the moment, the line shape of the absorbing sample mixes them to produce a 100-kHz signal

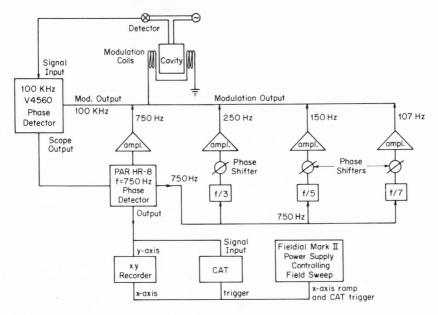

Fig. 1. Schematic diagram of experimental setup.

amplitude modulated at 750 Hz. The first PSD demodulates this signal, leaving a dc signal with 750 Hz modulation on it. This dc signal is, of course, the first derivative of the absorption line. The second PSD demodulates the 750-Hz component to yield the second derivative of the absorption line as a dc signal. The odd-subharmonic field modulation produces not only its own modulation frequency at the detector crystal, but also its harmonics in the proportion that they are represented in the absorption line shape. These frequencies are, similarly, mixed with the 100-kHz signal by the absorption line shape and are demodulated by the first PSD, but only their 750-Hz component is demodulated by the second PSD. These components correspond, of course, to the fourth, sixth, and eighth derivatives of the absorption line shape for the 750/3, 750/5, and 750/7 modulation frequencies, respectively. Thus the even derivatives of the original EPR absorption line appear added together in the output of the second PSD. When the phases and amplitudes of these derivatives are adjusted properly the result is a sharpened second-derivative presentation of the original absorption line. Adjusting all of these phases and amplitudes simultaneously is a very tedious business unless one has a strong signal such that scope presentation is possible during adjustment.

An alternative procedure was developed for achieving sharpened second-derivative EPR spectra without the necessity for simultaneous adjustment of the phases and amplitudes of the subharmonic modulation frequencies. An added bonus was a large improvement in signal-to-noise

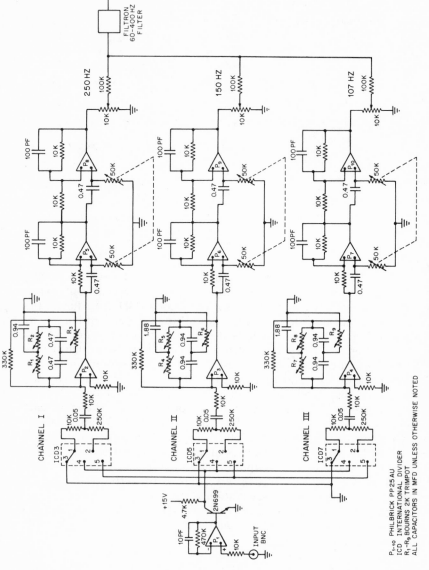

Fig. 2. Schematic diagram of odd-subharmonic generator.

ratio. The method utilizes a CAT to first record the second-derivative signal only, then the 750-Hz modulation is turned off, the 250-Hz modulation is turned on, and its phase and amplitude are adjusted by observing on a recorder the fourth-derivative signal. The fourth-derivative signal is then added to the second-derivative signal which is already in the memory of the CAT, and so on, until the sharpening has progressed to a satisfactory point as judged by the operator. The sharpening process is observed all the while by the operator, and intermediate stages of the process can be recorded as desired without erasing the memory of the CAT. The signal-to-noise ratio can be improved further by multiple passes for each of the derivative terms.

Figure 1 is self-explanatory, except for the amplifiers shown in the subharmonic channels, which are really one-half of an Eico Cortina stereo amplifier. The individual gain controls for the subharmonic channels are in the odd-subharmonic generator, and cross-talk between the subharmonic channels is eliminated by a resistance network at the common output of the device. The 750-Hz channel utilizes the other half of the stereo amplifier. The output transistors of the two halves of the stereo amplifier are protected from one another by 1-ohm resistors in series with each half. The V-4560 has its own protection for secondary modulation frequencies built in.

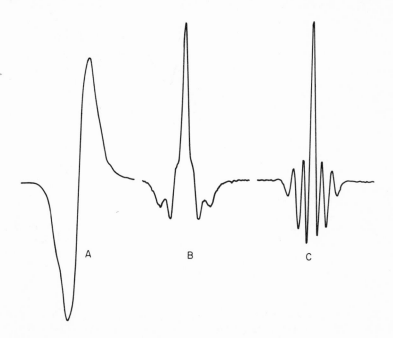

Fig. 3. Saturated solution of DPPH in benzene. (A) first derivative, (B) second derivative, (C) second plus fourth derivatives.

ODD-SUBHARMONIC GENERATOR

Figure 2 is a schematic of the odd-subharmonic generator used in the system. The square-wave reference at the modulation frequency of the PAR HR-8 is not large enough or fast enough to trigger the divider circuits, so an input circuit to shape and amplify the signal is used. The operational amplifier in the input amplifies the square wave from the PAR and adds to it the derivative of the square wave. The following transistor inverts the signal and removes the positive going pulse. Frequency division is obtained by integrated circuits labeled ICD-3, ICD-5, and ICD-7.* The resulting subharmonics are rectangular waves and their fundamental frequencies are extracted by active notch filters constructed by placing notch reject filters at the appropriate frequency in the feedback loop of an operational amplifier. The next stage

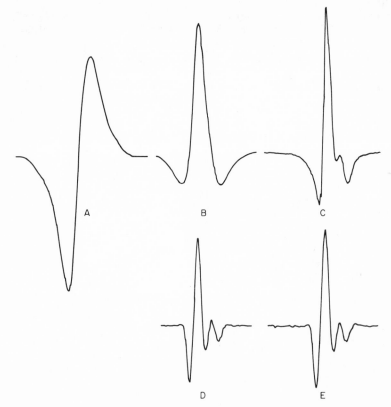

Fig. 4. Polycrystalline DPPH. (A) first derivative, (B) second derivative, (C) second plus fourth derivatives, (D) second plus fourth plus sixth derivatives, (E) second plus fourth plus sixth plus eighth derivatives.

*Obtained from the International Crystal Manufacturing Co., Inc., Oklahoma City, Oklahoma.

in the device is a unity-gain phase shifter[6] which provides greater than 270° of continuous phase shift. The final potentiometer in each channel is the gain control. The Filtron filter removes the high-frequency hash generated by the ICD trigger circuit.

RESULTS OF SECOND-DERIVATIVE LINE SHARPENING

The result of the second-derivative line sharpening is best demonstrated by sharpened spectra. Figure 3 shows the resolution enhancement obtained by going from first-derivative to second-derivative presentation, and the sharpening obtained by addition of one sharpening term for a saturated solution of DPPH in benzene. The first- and second-derivative EPR spectra of polycrystalline DPPH shown in Fig. 4 only hint at underlying structure; however, this structure is clearly resolved when two sharpening terms are added. Addition of the sharpening term from the eighth derivative offers no additional sharpening. The two peaks resolved are, of course, $g_{\parallel}$ and $g_{\perp}$ of

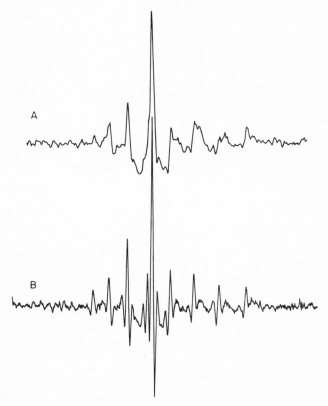

Fig. 5. (A) Second-derivative EPR spectrum of vanadyl *meso*-tetra-9-anthryl porphyrin in solid napthalene, (B) sharpened second-derivative EPR spectrum of this sample.

DPPH. All of these spectra were taken with the simultaneous addition method, which is satisfactory for strong signals. For weak signals the loss of 75 to 100 in signal-to-noise ratio for each sharpening term is often unacceptable. The spectrum in Fig. 5 of VO *meso*-tetra-9-anthryl porphyrin* was sharpened by adding about three passes each of the second, fourth, sixth, and eighth derivatives into the CAT. The signal-to-noise ratios of the unsharpened and the sharpened spectra are comparable within a factor of two.

CONCLUSION

The second-derivative line-sharpening device in conjunction with a CAT was found to give linewidth reductions of up to 50% with a relatively small amount of loss in signal-to-noise ratio compared to the simultaneous addition method. The odd-subharmonic generator is inexpensive to construct and is the only expense required for a laboratory already equipped with second-derivative equipment and a CAT.

ACKNOWLEDGMENTS

The author takes pleasure in thanking Dennis Wisnosky and Hans Tuche for many helpful contributions during the construction of the odd-subharmonic generator. In addition, the author wishes to thank the National Science Foundation for the Institutional Equipment Grant No. 206, which made construction of this instrument possible.

REFERENCES

1. Court L. Wolfe, Edmund C. Tynan, and Teh Fu Yen, unpublished work.
2. A. Hedberg and A. Ehrenberg, *J. Chem. Phys.* **48**, 4822 (1968).
3. Leland C. Allen, H. M. Gladney, and S. H. Glarum, *J. Chem. Phys.* **40**, 3135 (1964).
4. Sivert H. Glarum, *Rev. Sci. Instr.* **36**, 771 (1965).
5. Charles P. Poole, Jr., *Electron Spin Resonance* (Interscience, New York 1967).
6. Gene E. Tobey, *Electronic Design* **4**, 114 (February 15, 1967).

*Supplied by T. F. Yen, Mellon Institute.

ESR Study of Copper(II) and Silver(II) Tetraphenylporphyrin

P. T. Manoharan and Max T. Rogers

Department of Chemistry, Michigan State University
East Lansing, Michigan

Copper(II) tetraphenylporphyrin and silver(II) tetraphenylporphyrin have been studied by ESR spectroscopy using the pure solid materials, solutions, magnetically-diluted single crystals, and polycrystalline powders. It has been possible to obtain rather complete ESR data, including the hyperfine interaction tensors, for ^{14}N, ^{63,65}Cu, and ^{107,109}Ag. The results have been used to estimate molecular orbital coefficients in these compounds and to discuss details of the chemical bonding.

INTRODUCTION

The phthalocyanines and porphyrins consist of large, planar, conjugated ring systems which serve as tetradendate ligands; metallic cations can be easily accommodated at the center of these systems with the four nitrogens as the ligating atoms. The metalloporphyrins and metallophthalocyanines have been the subject of a considerable number of investigations in the past decade. A survey of the phthalocyanines has been done by Lever[1] and one on porphyrins appears in a book by Falk.[2] Figure 1 shows the molecular structure of the free base, tetraphenylporphyrin (H_2TPP).

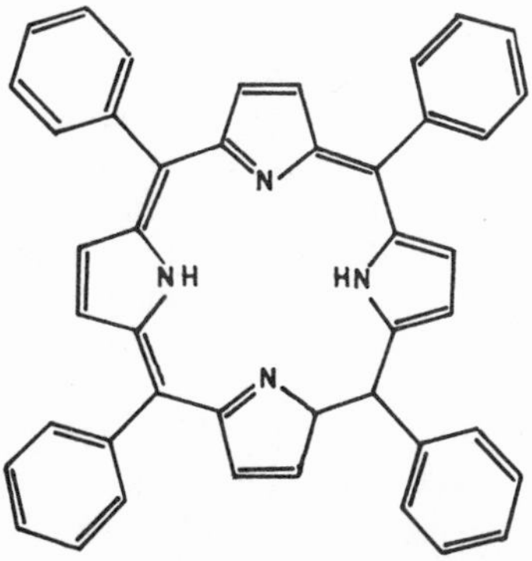

Fig. 1. The Structure of tetraphenylporphyrin.

In the present work we are mainly concerned with the electron spin resonance investigation of the tetraphenylporphyrin complexes of copper and silver in order to study the metal–nitrogen bonding. Ingram *et al.*[3] first investigated the paramagnetic resonance absorption spectra of the copper complex of $\alpha,\beta,\gamma,\delta$-tetraphenylporphyrin (CuTPP) and its *p*-chloro derivative. Their observation of hyperfine components, even in the concentrated crystals, led them to believe that the dipole–dipole interaction between the neighboring copper atoms in such large molecules is very much reduced. Moreover, the hyperfine pattern of the *p*-chloro-derivative, they thought, showed considerable interaction with the chlorine nuclei. This would imply that there is considerable overlap between the unpaired electron of the Cu^{2+} ion and the atomic orbitals of the peripheral chlorines, which are separated by at least 9 Å from the central copper atom. They interpreted this result as arising from an intermolecular interaction between the magnetic electron and the peripheral chlorine atoms *via* the π orbitals of the conjugated ring system. In other words, the phenyl groups were thought to be electronically coupled to the entire aromatic resonating system of the H_2TPP molecule. However, quite contrary to the above ESR conclusion, x-ray structural studies of H_2TPP and its metal derivatives[4–9] reveal that the phenyl rings make an angle of 60–90° with the plane of the porphyrin ring. Griffith,[9] on the basis of ESR and magnetic susceptibility measurements on copper, nickel, cobalt, and iron porphyrins and phthalocyanines, has proposed the following energy-level ordering for the $3d$ orbitals: $3d_{xy} < 3d_{xz,yz} < 3d_{x^2-y^2}$, with a variable position for $3d_{z^2}$. Koski and co-workers[10–12] studied the ESR spectra of the metal porphyrins in solutions and frozen glasses. For the copper and silver porphyrins they observed not only the metal hyperfine splittings, but also the ligand superhyperfine splittings due to the interaction of the unpaired electron with the nitrogen atoms of the pyrrole rings, and they calculated the bonding parameters for these complexes. Kivelson and Lee[13] studied vanadyl tetraphenylporphyrin by the ESR method and observed the nitrogen extrahyperfine structure, which they found to be very isotropic, indicating that the b_{2g}^* orbital is localized on the vanadium $3d$ orbital and that the in-plane π bonding is very slight. McGragh *et al.*[14] found that the solvents, the peripheral substituents on the porphyrin ring, and the temperature modify the spin resonance spectra of copper and silver porphyrins. They interpreted the variation of the *g*-factors and the nitrogen hyperfine splittings as resulting from the different states of solvation and solute aggregation that occur in these systems. Assour[15] studied the vanadyl, copper, and cobalt tetraphenylporphyrin complexes in solution and in diluted polycrystalline samples containing the diamagnetic free base, H_2TPP. His studies revealed distorted crystal-field surroundings which are more pronounced in the cobalt derivative than in any other. For copper complexes he concluded that in the solid the nonplanarity of the porphin nucleus has little influence on the ESR data, and that the unpaired electron is coupled equally to the four pyrrole nitrogens. His analysis also indicated a strong in-plane σ bonding characteristic of organometallic square-bonded complexes and little or no in-plane π bonding. He further noted that the out-of-plane π bonding was more significant in the vanadyl and cobalt derivatives than in the copper complex.

Though a fairly accurate account of the spin-Hamiltonian parameters can be obtained from a combined ESR study using the solution, the frozen-glass, and a polycrystalline sample diluted in a host lattice, a definitely clearer picture will only be obtained from single-crystal studies. Moreover, a detailed resolution of the superhyperfine lines in the ESR spectra of some polycrystalline samples may be possible if the studies are carried out in a Q-band instrument which will separate out the parallel and perpendicular features. The present study has been carried out to obtain a clear picture of bonding properties in copper and silver tetraphenylporphyrin (CuTPP and AgTPP) complexes. ESR study of these complexes has been carried out in both X-band and Q-band ESR spectrometers. This study allows us to compare the delocalization of the magnetic unpaired electron in the $3d$ and $4d$ atomic orbitals, respectively, of copper and silver. Also, a comparison of the dipole–dipole interactions in copper and silver complexes is possible from a study of the magnetically-concentrated species.

A comparison of bonding in the corresponding phthalocyanine complexes will also be attempted. ESR of copper phthalocyanine has been studied by Griffith,[9] Roberts and Koski,[16] Assour and Harrison,[17] Neiman and Kivelson,[18] Ingram and co-workers,[19,20] Deal et al.,[21] Assour and Kahn,[22] and Abkowitz et al.[23]

EXPERIMENTAL

The free base, H_2TPP, was kindly supplied to us by Dr. A. Adler of the New England Institute for Medical Research. The metallotetraphenyl-porphyrins were carefully prepared by the method of Dorough et al.[24] The tetraphenylporphyrin chelates were purified by repeated crystallization.

The free base H_2TPP was mixed with 0.5% of the appropriate paramagnetic chelate and dissolved in a mixture of benzene and xylene having the volume ratio 1:9. The solution was filtered and allowed to evaporate very slowly in a small beaker kept in the dark. Crystals of about $3 \times 3 \times 3$ mm size with very well-defined faces were obtained after about five to eight weeks.

All the ESR measurements were carried out using Varian V-4500-10A X-band and Q-band spectrometers with 100-kHz magnetic field modulation. The magnetic field in the X-band spectrometer was measured by determination of the NMR frequency with a Hewlett-Packard counter, while in the Q-band spectrometer the Varian Mark II Fieldial was used.

CRYSTAL STRUCTURES

Since the free base, tetraphenylporphyrin, has been used as host lattice for the ESR study of Cu^{2+} and Ag^{2+} complexes in this work, we must summarize the available information, not only on the crystal structures of

H_2TPP, but also on their metal derivatives. Though structure determinations have shown that tetraphenylporphyrin can be grown in monoclinic,[25] tetragonal,[5] orthorhombic,[26] and triclinic[6] modifications, the most common and readily crystallizable form is the triclinic one. Also, only for the triclinic modification has the relative orientation of the molecular and crystallographic axes been reported.[6] The unimolecular unit cell, with dimensions $a = 6.44$, $b = 10.42$, $c = 12.41$ Å, $\alpha = 96.06$, $\beta = 99.14$ and $\gamma = 101.12°$, is centrosymmetrical, with two independent pyrrole and phenyl groups. The nonplanarity of the porphyrin ring is evidenced by the finding that one pair of centrosymmetrical pyrroles is essentially coplanar with the nuclear-least-squares plane of the porphyrin ring, and the other pair carrying the central hydrogen atoms is inclined $\pm 6.6°$ to this plane. The phenyl groups are rotated out of the plane of the porphyrin ring by about 60°. Also, the atoms of the inner 16-member ring are indirectly planar, though the porphyrin ring is nonplanar. The most interesting points of the structure[6] are the presence of two sets of nitrogen pairs and the isolation of the phenyl groups electronically from the inner ring system. The packing mode of the pyrrole groups of the adjacent molecules is such that two opposite pairs of pyrrole groups (designated N-1 and N-2) lie, respectively, perpendicularly above an N-2 and N-1 group of molecules below. The closest intermolecular contact between the two pyrrole groups of adjacent molecules is 3.63 Å.

In the case of CuTPP, although the copper atom is surrounded by four nitrogens, the nitrogens are not coplanar.[4] Also, CuTPP has a tetragonal unit cell with the $Cu-N$ distance < 2.05 Å and the four nitrogens noncoplanar. The deviation from planarity seems to be greater in CuTPP than in H_2TPP itself. It is also evident from the many investigations mentioned here that the deviations from planarity are different for each chelate, depending on the crystallographic packing of the molecules in each case. Not much structural work has been done on AgTPP. Tulinsky[27] has found that AgTPP crystallizes the same way as the free base in a triclinic modification. Also important is the fact that H_2TPP and AgTPP are isostructural, with space group $P\bar{1}$ and with one molecule per unit cell. It is evident from the work of Donnay and Storm[8] that most solid solutions of AgTPP/H_2TPP are triclinic. Particular mention may be made of the observation by Donnay and Storm[8] that two very-different-looking crystals crystallize from one and the same solution.

RESULTS

Magnetically-Concentrated Chelates

Ingram et al.[3] were the first to observe the metal hyperfine splittings in magnetically-concentrated crystals of CuTPP. Surprisingly, the ESR spectra of VOTPP and CoTPP also display such hyperfine splittings in magnetically-concentrated form.[4] We repeated these experiments for CuTPP and found that the parallel components of the hyperfine splittings were easily resolved in the Q-band ESR spectrum of this metal chelate, as shown in Fig. 2a, but

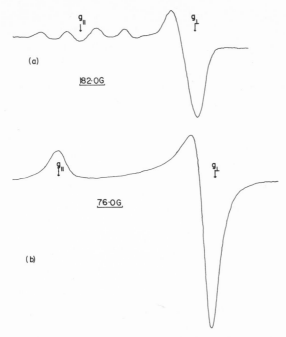

Fig. 2. First-derivative ESR spectra of polycrystalline samples
of magnetically-pure (a) CuTPP and (b) AgTPP at 298°K
(Q-band).

the perpendicular components were not resolved. As was explained several
times earlier, the appearance of the metal hyperfine lines can be attributed
to a considerable decrease in the dipolar interactions. This could be achieved
only by such physical stacking of the paramagnetic TPP molecules that the
out-of-plane phenyl rings might shield the paramagnetic ion from its nearest
neighbors. This would effectively reduce the magnetic interaction in the
crystal.

Though such resolutions of hyperfine lines have been observed in
VOTPP, CoTPP, and CuTPP, we failed to observe any resolution of metal
hyperfine lines in pure AgTPP. The Q-band ESR spectrum of pure AgTPP
powder is shown in Fig. 2b. The only resolutions we could observe were
those of the Zeeman splittings corresponding to the $g_{\parallel}$ and $g_{\perp}$ tensors. Even
at liquid-nitrogen temperature, no hyperfine splittings due to [107,109]Ag
could be observed. This new phenomenon unknown to the TPP chelates
requires an additional explanation of the physical stacking of the nonplanar
phenyl rings. The nearest Cu—Cu distance in the tetragonal unit cell of
CuTPP is 8.2 Å, while the nearest Ag—Ag distance in the unit cell of AgTPP,
which is isostructural with the free base H_2TPP, would be about 3.2 Å. In
other words, the dipolar interaction between the paramagnetic ions would

be greater in AgTPP than in CuTPP. Moreover, the structure of CuTPP[7] is such that the metal ion is effectively shielded by benzene rings from the neighboring molecules in both axial positions. Such a protective diamagnetic shielding is absent in the AgTPP structure, where the nearest Ag^{2+} ions are stacked almost one above the other. In such a case effective dipolar inter-action will broaden the ESR lines. The ESR spin-Hamiltonian parameters obtained from concentrated crystals appear in Table I.

TABLE I
ESR Parameters in Magnetically-Pure Powders[a]

Complex	$g_{\parallel}$	$g_{\perp}$	A	B
CuTPP	2.179 (2.17)	2.033 (2.05)	212.2 (250)	(30)
AgTPP	2.110	2.030	—	—

[a]Hyperfine splittings are in $cm^{-1} \times 10^{-4}$. Values in parentheses are from Ingram et al. [3]

Solutions

The room-temperature ESR spectra of 5×10^{-3} M solutions of CuTPP and AgTPP in benzene are shown, respectively, in Figs. 3a and 3b. Our spectra for CuTPP in benzene and chloroform resemble those for CuTPP in $CHCl_3$ reported by Assour,[15] and consist of four resolved copper lines. The magnetic interaction of the unpaired electron with the pyrrole nitrogens further splits the two high-field copper lines. The nine superhyperfine components of each of the high-field copper lines have equal separations of 16.07 G, and intensity ratios $1:4:10:16:19:16:10:4:1$. The isotropic spin-Hamiltonian parameters were determined using the equation

$$\mathcal{H}_{sp} = \langle g \rangle \beta H_z S_z + \langle a \rangle SI + \sum_N \langle a_N \rangle SI^{(N)} \tag{1}$$

where $\langle g \rangle = \frac{1}{3}(g_{\parallel} + 2g_{\perp})$; $\langle a \rangle = \frac{1}{3}(A + 2B)$, with A and B the hyperfine splittings of the metal nuclei, $^{63,65}Cu$ and $^{107,109}Ag$; a_N is the isotropic superhyperfine splitting due to the pyrrole ring nitrogen atoms (^{14}N); I and $I^{(N)}$ refer to the nuclear spin of the metal and ligand nitrogen nuclei, respec-tively, and the last term is to be summed over the four nitrogen nuclei. Experimental distances between the adjacent copper resonance lines being unequal, we applied second-order perturbation theory to get the average spin-Hamiltonian parameters.[28] The second-order equation

$$H_m = H_0 - am - (a^2/2H_0)[I(I + 1) - m^2] \tag{2}$$

was used, where $H_0 = h\nu/\langle g \rangle \beta$, m is the nuclear spin quantum number identified with each resonance line, and H_m is the resonance value of the applied field for this line. For CuTPP in benzene solution we obtained the values $\langle g \rangle = 2.0842 \pm 0.0002$, $\langle a \rangle = 92.00 \pm 0.5$ G, and $\langle a_N \rangle = 16.07$ G.

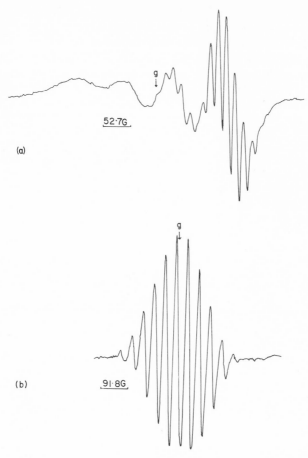

Fig. 3. First-derivative ESR spectra of 0.005 M solutions of (a)
CuTPP, and (b) AgTPP, in benzene at 298°K (X-band).

The $\langle g \rangle$ and $\langle a \rangle$ values reported by Assour[15] for CuTPP in CHCl$_3$ solution
differ slightly from our measured values for the chloroform solution. This
difference in values for benzene and chloroform solutions must be due to a
solvent effect, but the isotropic superhyperfine splitting due to the pyrrole
nitrogens does not seem to vary much. The splittings due to ^{63}Cu and ^{65}Cu
were not resolved because of line broadening.

The room-temperature spectrum of AgTPP (Fig. 3b) in benzene is a very
interesting one. It consists of 11 almost equidistant lines, the line separations
being a little smaller at the center than at the edges. First-order analysis of
this spectrum reveals that the distance between the fifth and the seventh line
is approximately the hyperfine splitting from ^{107,109}Ag $(I = \frac{1}{2})$, while the

line separations at the extreme ends of this spectrum are probably the right order of magnitude to be the superhyperfine splittings from ^{14}N of the pyrrole groups. It is worth mentioning at this point that the intensities of the eleven lines have approximately the ratios $1:4:11:20:29:32:29:20:11:4:1$, indicating that two sets of nine lines, each from four planar nitrogens and with intensity ratios $1:4:10:16:19:16:10:4:1$, have overlapped. We did not resolve the splittings from ^{107}Ag and ^{109}Ag, although the magnetic moments differ by about 15%. Analysis of the spectra based on the spin-Hamiltonian of Eq. (1), with no second-order corrections, yields $\langle g \rangle = 2.0603 \pm 0.0005$, $\langle a \rangle = 42.73 \pm 0.5$ G, and $\langle a_N \rangle = 22.93 \pm 0.10$ G. Values obtained from spectra in chloroform solution differ slightly from those obtained from benzene-solution spectra. The ESR measurements on a 10^{-3} M pyridine solution of AgTPP gave the following spin-Hamiltonian parameters: $\langle g \rangle = 2.0585 \pm 0.0004$, $\langle a \rangle = 43.00 \pm 0.7$ G, and $\langle a_N \rangle = 23.48 \pm 0.1$ G. All the results are compiled in Table II.

Frozen Glass

The ESR spectrum of the frozen solution of CuTPP is composed of two sets of metal hyperfine lines corresponding to the $g_{\parallel}$ and $g_{\perp}$ tensors. The metal hyperfine lines are further split by the superhyperfine interaction with the nitrogens of the pyrrole group. This ESR spectrum was interpreted with the spin Hamiltonian for axial symmetry,[29] using the coordinate system shown in Fig. 4:

$$\mathscr{H}_{sp} = g_{\parallel}\beta H_z S_z + g_{\perp}\beta(H_x S_x + H_y S_y) + A S_z I_z + B(S_x I_x + S_y I_y)$$
$$+ Q'[I_z^2 - \tfrac{1}{3}I(I+1)] + \sum_n S A_n I_n \qquad (3)$$

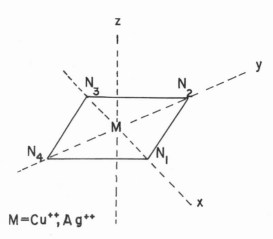

Fig. 4. The coordinate system for CuTPP and AgTPP molecules.

TABLE II
Spin-Hamiltonian Parameters from Solutions and Frozen Glasses[a]

Complex	Solvent	$\langle g \rangle$	$\langle a \rangle$	$\langle a_N \rangle$	$g_{\parallel}$	$g_{\perp}$	A	B	A_N	B_N	Q'	
CuTPP	Chloroform	2.0868 ±0.0005	88.10 ±0.4	15.66 ±0.005	2.187 ±.002	2.032 ±0.003	209 ± 3	31.8 ±0.5	14.48 ±0.05	15.9 ±0.3	≤4.0	I[b]
					2.183 ±0.002	2.027 ±0.003	209 ± 3	31.7 ±0.5	14.45 ±0.02	15.9 ±0.3	≤4.0	II[b]
	Benzene	2.0842 ±0.0002	89.5 ±0.5	15.63 ±0.05	—	—	—	—	—	—	—	
	Pyridine	2.1070 ±0.0005	88.57 ±0.6	15.35 ±0.05	—	—	—	—	—	—	—	
AgTPP	Chloroform	2.0592 ±0.0005	41.55 ±0.10	22.02 ±0.05	2.109 ±0.003	2.038 ±0.004	(c)	(c)	(c)	(c)	—	
	Benzene	2.0603 ±0.0005	41.1 ±0.5	22.06 ±0.1	—	—	—	—	—	—	—	
	Pyridine	2.0585 ±0.0004	41.3 ±0.7	22.6 ±0.1	—	—	—	—	—	—	—	

[a] Hyperfine values and Q' are given in $cm^{-1} \times 10^{-4}$.
[b] The two sets of parameters I and II belong to the two different types of molecule present.
[c] Could not be determined precisely.

The fifth term in this expression is the quadrupolar term and the last term describes the superhyperfine interaction, summed over the four nitrogen nuclei. That term will be discussed in detail later. The quadrupolar term is not applicable to the AgTPP complex because $I = \frac{1}{2}$ for ^{107,109}Ag.

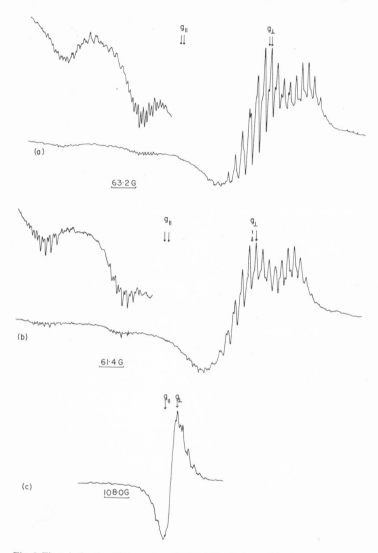

Fig. 5. First-derivative ESR spectra of 0.0035 M solutions of (a) CuTPP at 77°K, (b) CuTPP at 153°K, and (c) AgTPP at 77°K, in chloroform (X-band). Note that the first two sets of low-field lines from the parallel orientation of CuTPP are amplified to clearly show the change in intensity ratios of the lines due to the two species.

The ESR spectrum of 0.0035 M CuTPP in chloroform, measured at 77°K, is shown in Fig. 5a. It is similar to that reported by Assour,[15] but the resolution of the perpendicular component is better, thus permitting more precise measurements of $g_\perp$, $B(^{63,65}Cu)$, and the nitrogen hyperfine splitting. The low-field copper lines with $m = \frac{3}{2}, \frac{1}{2}$, and $-\frac{1}{2}$ are clearly resolved and each is split into 18 narrow lines with separations approximately 7.1 G. Assuming that ^{63}Cu and ^{65}Cu lines are not resolved, one would expect only nine components. Assour[15] accounted for the extra nine lines by postulating that in frozen solutions there are two different Cu^{2+} sites with different $g_{\parallel}$ values. This assumption was justified by the following observations: (1) the observed hyperfine lines (18 on each Cu line) appear to alternate in their intensities, which allows them to be grouped into two sets with equal spacings of 14.4 G; (2) the ratios of the intensities for the more intense set seem to be in good agreement with those expected for four equivalent N nuclei; (3) the spectra of polycrystalline solid CuTPP doped in free H_2TPP which he obtained show only nine equally-spaced superhyperfine lines with 14 G separations, suggesting that four nitrogens are involved and there is no resolution of ^{63}Cu and ^{65}Cu lines; and (4) the flexibility of the porphyrin skeleton and its ease of adaptability to the crystalline environment might very well give rise to two different Cu^{2+} centers in frozen solutions; CoTPP, as an example, seems to have three such centers. We agree with this assignment, and have additional evidence in favor of it. If the two sets resulted from the resolution of the ^{63}Cu and ^{65}Cu isotopes (69.09% and 30.91% natural abundance), their intensities should always have a ratio about 2:1. We find, however, that the intensity ratios depend on temperature and on the rate of cooling of the sample. When the temperature is lowered gradually from 190°K to 120°K by flowing cooled nitrogen gas around the sample the intensity ratios are nearly 2:1, but vary somewhat (Fig. 5b). When the sample is cooled quickly by placing it directly into liquid nitrogen the intensity ratios become about 1:1. We conclude that two species are present, the relative amounts depending on conditions. It is not correct to assume that the ^{63}Cu and ^{65}Cu lines cannot be resolved, since we have been able to do this, at least for the parallel component in polycrystalline powder and single-crystal measurements. The complete results of our analysis of these spectra are given in Table II.

The frozen solution of AgTPP (0.003 M) in $CHCl_3$ gave the ESR spectrum shown in Fig. 5c. The spectrum is composed of Zeeman splittings due to $g_{\parallel}$ and $g_\perp$ which are further split by the hyperfine interactions with $^{107,109}Ag$ and the superhyperfine interactions with four planar nitrogen atoms. Unfortunately, the analysis of the spectrum is made complicated by the poor resolution of the hyperfine lines and by the considerable overlapping of the resonance transitions due to the parallel and perpendicular components. The only things that are easily visible are the intense lines associated with $g_\perp$; the hyperfine lines associated with $g_{\parallel}$ are poorly resolved. The values for the spin-Hamiltonian parameters of Eq. (3) cannot be obtained from this spectrum with any reasonable accuracy. However, we can say with certainty that there is no resolution of ^{107}Ag and ^{109}Ag in either of the axially symmetrical tensors. Evidence is also completely lacking for the presence of more than one Ag^{2+} center in the frozen solution. The less-intense lines are due to the parallel component of the spectrum.

Polycrystalline Samples

Single crystals of free base H$_2$TPP doped with small (0.5%) admixtures of the paramagnetic materials were grown from solutions in a xylene–benzene mixture. After studying these single crystals by the ESR method (see below) they were powdered and the polycrystalline samples studied both at X-band and at Q-band. The measurements were taken at room temperature and also at liquid-nitrogen temperature, using liquid nitrogen as coolant. The spectra in the two cases were identical, so we report here only the spectra taken at room temperature.

The first-derivative ESR spectra of CuTPP diluted in H$_2$TPP and measured at X-band and Q-band are shown in Figs. 6a and 6b, respectively. The spin-Hamiltonian parameters were obtained with the help of Eq. (3). The X-band spectrum resembles the one reported by Assour[15] and consists of two well-resolved and easily-identifiable hyperfine components of the copper isotopes. Along the parallel direction the hyperfine components overlap considerably with those of the perpendicular hyperfine lines. Each one of the parallel hyperfine lines has the hyperfine splittings due to ^{63}Cu and ^{65}Cu well resolved. This is the only difference between our spectra and the one reported by Assour.[15] Furthermore, each of these copper hyperfine components are further split into nine lines of equal separation (14.2 G) with the intensity ratios 1:4:10:16:19:16:10:4:1. However, the overlapping of the hyperfine lines in the high-field region makes it difficult to measure the spin-Hamiltonian parameters of the perpendicular components.

Fortunately, the ESR measurement of this sample at Q-band (Fig. 6b) clearly separates the parallel and perpendicular components. The parallel part is composed of four resonance transitions due to the copper isotopes. Each one of these transitions further gives rise to nine superhyperfine lines with equal separations, resulting from the interaction of the magnetic electron with the four ^{14}N nuclei from the pyrrole groups. More important is the observation that each of these four parallel transitions is actually composed of 18 lines, the alternate lines being about half as intense as the others. This proves the resolution of lines from interaction with the ^{63}Cu and ^{65}Cu isotopes, since the natural abundance of ^{63}Cu is slightly more than twice that of ^{65}Cu. However, a careful analysis showed that the nitrogen hyperfine constants remain constant at 14.2 G regardless of whether the lines from ^{63}Cu or from ^{65}Cu are examined. This enabled us to calculate the splitting constants for ^{63}Cu and ^{65}Cu separately, and it was found that the splittings so obtained were in the ratio of the magnetic moments of these two nuclei within experimental error.

The perpendicular part of the spectrum now becomes fairly simple to analyze. It is composed of 15 equidistant lines with separations of 16.91 G.

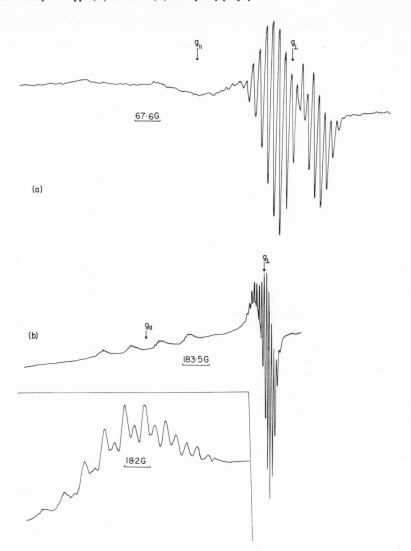

Fig. 6. First-derivative ESR spectra of polycrystalline CuTPP doped in the free base H_2TPP at room temperature (a) in X-band and (b) in Q-band. Note that the insert in (b) is obtained by increasing the gain for the set of low-field lines to demonstrate the separation of ^{63}Cu and ^{65}Cu hyperfine lines.

Two such separations will be equal to the hyperfine splittings of $^{63,65}Cu$ along the perpendicular direction. No resolution of the ^{63}Cu and ^{65}Cu isotopes in this set is observed. The clear separations of $g_{\parallel}$ and $g_{\perp}$ parts in the Q-band spectrum enabled us to calculate the spin-Hamiltonian parameters

TABLE III
Spin-Hamiltonian Parameters from Polycrystalline and Single-Crystal Measurements[a]

Sample	Measurement	$g_\parallel$	$g_\perp$	A	B	A_N	B_N	Q'
CuTPP/H$_2$TPP	Single-crystal	2.190 ±0.002	2.045 ±0.002	210.9 (^{65}Cu) ±0.9 201.3 (^{63}Cu) ±1.0	33.03 ±0.2	14.56 ±0.03	16.14 ±0.03	≤4.0
	Polycrystalline	2.187 ±0.003	2.045 ±0.003	214.1 (^{65}Cu) ±1.0 202 ± 1.1 (^{63}Cu)	32.85 ±0.3	14.5 ±0.05	16.1 ±0.03	≤4.0
AgTPP/H$_2$TPP	Single-crystal	2.108 ±0.003	2.037 ±0.0005	56.62 ±0.5	28.28 ±0.3	20.46 ±0.5	24.05 ±0.5	—
	Polycrystalline	2.107 ±0.003	2.032 ±0.001	56.7 ±0.5	28.50 ±0.5	20.90 ±0.3	24.05 ±0.5	—

[a]Hyperfine values and Q' are given in cm$^{-1} \times 10^{-4}$.

accurately. The values are given in Table III. Some values are slightly different from those reported by Assour.[15]

At least two important factors are evident from this study. The cupric ion substitutes in the porphyrin ring in only one way in crystals. Also, the nonplanarity observed in x-ray structural work is not manifested in ESR work, or else the four nitrogens of the pyrrole group are planar (or nonplanar with only very slight deviations). This is evident from our observation that the nitrogen resonance lines are equally spaced in both parallel and perpendicular components of the spectrum.

The X-band and Q-band ESR spectra of polycrystalline AgTPP diluted in H_2TPP and measured at room temperature are shown in Figs. 7a and 7b, respectively. The X-band spectrum shows a set of less-intense lines in the

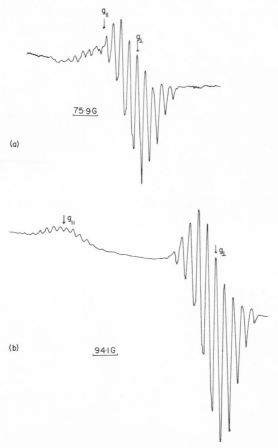

Fig. 7. First-derivative ESR spectra of polycrystalline AgTPP doped in the free base H_2TPP at room temperature (a) in X-band and (b) in Q-band.

low-field region of spectrum, merging later into the intense high-field lines. The less-intense lines are due to the $g_{\parallel}$ part, composed of resonance transitions due to ^{107,109}Ag and the four pyrrole nitrogens. Similarly, the ten high-field, equally-spaced lines originate from the $g_{\perp}$ component. Using Eq. (3), one is able to analyze this spectrum, but with limited accuracy because of the overlapping of the $g_{\parallel}$ and $g_{\perp}$ lines. However, the Q-band spectrum (Fig. 6b) is definitely superior in resolution, and hence easier to analyze. The parallel part consists of 12 lines spaced almost equally except that the spacings at the center are a little smaller than the ones at the extreme ends. Our analysis indicates that roughly three such spacings comprise the hyperfine splittings by the ^{107,109}Ag nuclei. Detailed resolution of these lines failed to show the nuclear interaction of the $4d$ electron with ^{107}Ag and ^{109}Ag separately. However, the superhyperfine structure from interaction with the four nitrogens is seen, and averages out to 21.0 G.

The perpendicular part of the spectrum consists of ten equally-spaced lines with small satellites on each one of them. A careful analysis proves that $B(^{107,109}$Ag) has a value equal to slightly more than one of those equal spacings, while one such equal spacing is the superhyperfine splitting due to the ^{14}N nuclei. The spin-Hamiltonian parameters calculated from this spectrum are found in Table II.

This polycrystalline study confirms that, at least in the lattice of free base H_2TPP, Ag^{2+} does not occur in more than one type of ligand surrounding. The deviation from planarity, if any, has not been detected by the ESR work. As in the case of CuTPP, the nitrogen resonance lines are equally spaced, indicating the nearly-planar nature of the porphyrin nucleus. The flexibility of the porphyrin skeleton has been restricted by the presence of strongly-interacting metal ions. This is the fairest conclusion to which one can come after a detailed ESR study of the polycrystalline copper and silver tetraphenylporphyrins.

Single Crystals

Single crystals of free base H_2TPP were grown doped with about 0.5% of the paramagnetic impurity from solutions of xylene and benzene. Most of the crystals had well-developed faces. All the crystals obtained by this method were triclinic, as were those used in the crystal-structure study of Silvers and Tulinsky;[6] they were found to contain only one molecule per unit cell,[6] and hence there should be only one equivalent magnetic site in the crystal. However, when we rotate the crystal around an axis perpendicular to the plane containing the four nitrogens of one unit cell the adjacent nitrogen plane obtained after a 90° rotation will be found to be slightly different from the initial one.[6] This small difference is actually found to affect the measured single-crystal ESR spectra.

For ESR measurements at X-band the crystal was mounted in a two-circle goniometer crystal holder[30] constructed of Teflon and Kel-F. After mounting the crystal in a selected orientation it could then be rotated about two perpendicular directions to facilitate the search for maximum and minimum values of the hyperfine splittings. In crystals such as CuTPP and AgTPP this proves to be the best procedure; the fourfold axis of the molecule may be located directly, since the maximum hyperfine splittings by [63,65]Cu or [107,109]Ag are observed when the magnetic field lies along it. The coordinate system used for analysis of the g tensor and the hyperfine splitting tensors is shown in Fig. 4; the z axis is chosen to be the fourfold molecular axis. The metal hyperfine splittings were found to be a maximum along the fourfold axis of the molecule from the polycrystalline spectra, so the z axis was located and the angular variation of the spectra studied with the magnetic field in the plane containing the z axis. The crystal was then remounted with the z axis as the rotation axis and the angular variation of the spectra recorded with the magnetic field in the xy plane. These spectra are sufficient to provide the principal components of the g tensor and the hyperfine splitting tensors. The data also provide information concerning the planarity (or nonplanarity) of the porphyrin ring.

The ESR spectra from single-crystal studies may be interpreted in terms of the spin Hamiltonian of Eq. (3). The electronic Zeeman and metal hyperfine tensors can be tested with the equations

$$g^2 = g_z^2 \cos^2\theta + g_x^2 \sin^2\theta \cos^2\phi + g_y^2 \sin^2\theta \sin^2\phi \qquad (4)$$

and

$$g^2 A^2 = A_z^2 g_z^2 \cos^2\theta + A_x^2 g_x^2 \sin^2\theta \cos^3\phi + A_y^2 g_y^2 \sin^2\theta \sin^2\phi \qquad (5)$$

where the spherical polar angles θ and ϕ relate the external magnetic field H to the z and y axes, respectively, and g and A are the measured values for any specified θ and ϕ as defined by these equations. In case of complete square-planarity of the metal and the four nitrogen nuclei, g_z, $g_x = g_y$, A_z, and $A_x = A_y$ will correspond, respectively, to $g_\parallel$, $g_\perp$, A, and B of Eq. (3). Hence the fitting of the experimental data to the relations (4) and (5) must reveal any nonplanarity of the four nitrogens.

The ESR spectrum of CuTPP diluted in H_2TPP is shown in Fig. 8a when the applied static magnetic field is parallel to the fourfold molecular axis. It consists of four sets of lines due to the interaction of the unpaired electron with the copper nuclei ($I = \frac{3}{2}$). The four nitrogens of the pyrrole groups, if equivalent, would give rise to nine equidistant lines by interaction with the unpaired electron. The spectrum (Fig. 8a) shows that each of the four sets of copper lines consists of nine equidistant components interspaced in some cases by a second set of components with about one-half the intensities of those of the stronger set. The weaker components are resolved in the low-field regions of the first two sets of copper lines and in the high-field regions of the other two sets. This indicates that the interactions with [63]Cu and [65]Cu are being resolved, with the weaker components arising from interaction with [63]Cu (abundance 30.91% and magnetic moment slightly greater than [65]Cu). Failure to resolve these components in the high-field regions of the first two sets of copper lines and the low-field regions of the second two sets is attributed to the presence of two very slightly different

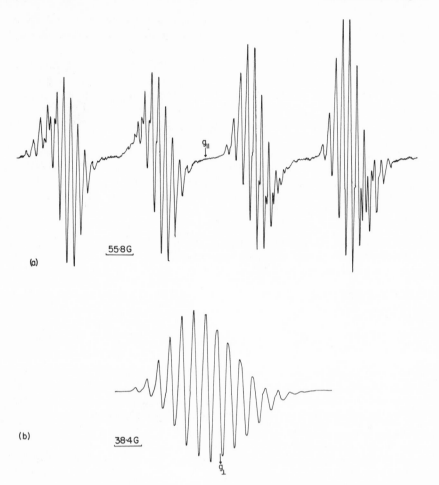

Fig. 8. First-derivative ESR spectra of a single crystal of CuTPP in H_2TPP at room temperature (a) when the applied static field is parallel to the z axis and (b) when the applied static field is in the xy plane and parallel to the line bisecting the x and y axes (X-band).

nitrogen planes as mentioned above for crystals of the free base H_2TPP. Two different spectra with slightly different g values would then overlap enough to prevent resolving the ^{63}Cu and ^{65}Cu in the single crystal. However, in the polycrystalline material these orientations must average to the two permitted orientations, giving rise to only the spectra of the parallel and perpendicular orientations. The Q-band spectrum of the polycrystalline material (Fig. 6b) shows the presence of a single Cu^{2+} center with the ^{63}Cu and ^{65}Cu lines resolved along the parallel orientation. The nitrogen hyperfine splitting in this orientation, 14.24 G, is the minimum value observed, while the hyperfine splittings due to ^{63}Cu and ^{65}Cu are maximal.

The angular variation of the spectra was then studied every 10°, or 2° when necessary, in the plane containing the z axis of the molecule. As the angle θ between the applied static field and the z axis increases by moving away from the z axis, the nuclear hyperfine splitting due to ^{63}Cu and ^{65}Cu not only decreases, but the lines merge together to form a single line. Also, the nitrogen hyperfine splitting increases with increase in θ. This trend continues until a maximum is reached for the nitrogen hyperfine splitting at $90 \pm 2°$ from the z axis. The spectrum obtained at this particular orientation is shown in Fig. 8b. Here the hyperfine structure collapses to a group of 15 almost equidistant lines spread over approximately 250 G. Actually, this spectrum is a superposition of four groups of nine lines arising from the interaction of the unpaired electron with the copper and four equivalent nitrogen nuclei. In this orientation the hyperfine lines due to ^{63}Cu and ^{65}Cu were not resolved, but the hyperfine splittings due to ^{63,65}Cu and the ^{14}N nuclei were analyzed easily. The analysis again indicates that there are four equivalent nitrogens. When θ is in the region near 90° a small quadrupolar effect is evident, but the quadrupolar coefficient Q' in Eq. (3) cannot be as accurately determined as the other parameters. The angular-variation study was further continued for another 90° until $\theta = 180°$, at which angle the recorded spectrum was similar to that at $\theta = 0°$. All these spectra at different angles comply with Eqs. (4) and (5), and ϕ was calculated to be 45° with $g_x = g_y = g_\perp$ and $A_x = A_y = B$.

The crystal was then remounted with the z axis as the rotation axis. The angular variation study of the ESR spectrum in the xy plane of this molecule reveals that there is only a slight variation of the spectra at different orientations. Moreover, neither the g value nor the hyperfine splitting constants vary appreciably. However, we find that the four nitrogens are not in an exact plane, but that one of the two opposed pairs of nitrogens is about 2 to 3° below the nuclear least-squares plane; this figure is about 6° in the host lattice itself.[6] All the spectra were analyzed with the aid of the spin Hamiltonian, Eq. (3), and the results are shown in Table III.

The ESR study of AgTPP in diluted single crystals is interesting in its own right. The measurements were made the same way as for CuTPP. The only difficulty we faced was to find the z axis of the molecule; this was located after many attempts. The ESR spectrum of this molecule at $\theta = 0$ is shown in Fig. 9a. The spectrum consists of twelve almost equidistant lines, the distances of separation being slightly less at the center than at the ends. This is actually a superposition of two sets of nine lines, the two sets arising from the interaction of the unpaired electron with the nuclei of the central metal ion Ag^{2+} and the four ligand nitrogens. The nine lines from each set must have intensity ratios $1:4:10:16:19:16:10:4:1$. It is very interesting to note that the hyperfine splitting due to the ^{107,109}Ag $(I = \frac{1}{2})$ isotopes is approximately the distance between the fifth and the eighth lines. Mention should be made of the fact that we could not see the separate lines from ^{107}Ag and ^{109}Ag. The calculations of the g value and the hyperfine values do not reveal any second-order effects. The angular variation of the spectrum

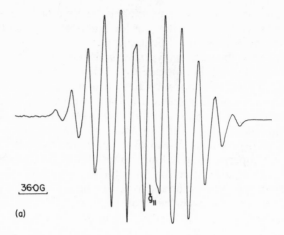

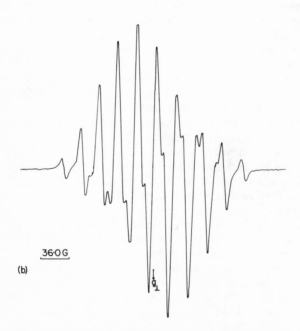

Fig. 9. First-derivative ESR spectra of a single crystal of
AgTPP in H$_2$TPP at room temperature (a) when the applied
field is parallel to the z axis and (b) when the applied field is
parallel to the line bisecting the x and y axes in the xy magnetic
plane (X-band).

shows that there is only one type of AgTPP molecule in the crystal and that all the four nitrogens are roughly equivalent. Moreover, the g value and the metal hyperfine values fit Eqs. (4) and (5). The ESR spectrum at $\theta = 90°$ is shown in Fig. 9b. It is formed again by the superposition of two sets of nine equidistant lines. The calculation of the g value and the hyperfine tensors was made in the usual manner. The ESR spin-Hamiltonian parameters obtained from the analysis of AgTPP spectra appear in Table III. Here again the general conclusion is that there is very little deviation from planarity of the nitrogen ligands and that the four nitrogens are equivalent. Also, there is only one type of AgTPP in the crystal lattice.

DISCUSSION

The ESR data can be interpreted in terms of molecular orbitals. Visualizing an ideal model for CuTPP and AgTPP molecules, they would show D_{4h} symmetry if all the four nitrogens and the metal atom form a square-planar arrangement. By group theory, the antibonding molecular orbitals necessary for our discussion can be obtained by linear combinations of the proper ligand functions and the $3d$ and $4d$ functions of copper and silver, respectively:

$$\psi(b_{1g}) = \beta_1 d_{x^2-y^2} - \beta_1'(\sigma_1 - \sigma_2 + \sigma_3 - \sigma_4)$$
$$\psi(a_{1g}) = \alpha d_{z^2} - \tfrac{1}{2}\alpha'(\sigma_1 + \sigma_2 + \sigma_3 + \sigma_4)$$
$$\psi(e_g) = \varepsilon d_{xz} - (1/\sqrt{2})\varepsilon'(p_1 - p_3)$$
$$= \varepsilon d_{yz} - (1/\sqrt{2})\varepsilon'(p_2 - p_4) \tag{6}$$
$$\psi(b_{2g}) = \beta_2 d_{xy} - \tfrac{1}{2}\beta_2'[p_1 - p_2 + p_3 - p_4]$$

The ligand functions are of the appropriate symmetry. In particular, the σ functions are obtained by the proper hybridization of $2s$ and $2p$ orbitals of the ligand nitrogens (see below). Griffith[9] has given the energy-level ordering of $3d$ orbitals in the phthalocyanines and porphyrins as $3d_{xy} < 3d_{xz,yz} < 3d_{x^2-y^2}$, with a variable position for $3d_{z^2}$. The position of $3d_{z^2}$ depends, of course, on the strength and type of axial distortion. Quite interestingly enough, for the cobalt and copper tetraphenylporphyrin complexes Zerner and Gouterman,[31] on the basis of an extended Hückel calculation, predicted the ordering to be $3d_{xy} < d_{xz,yz} < d_{z^2} \ll d_{x^2-y^2}$. Whatever the ordering of the $3d$ orbitals may be, the ground state of the copper and silver tetraphenylporphyrin complexes is definitely $^2B_{1g}$, with the unpaired electron in the nondegenerate b_{1g} molecular orbital. Accordingly, Eq. (6) shows the various states in order of decreasing energy. The b_{1g} and a_{1g} molecular orbitals account for the σ bonding, the e_g orbital for the out-of-plane π bonding, and the b_{2g} orbital for the in-plane π bonding.

The ESR spin-Hamiltonian parameters must throw light on the nature of the various bonding coefficients in Eq. (6). Using the treatment of Abragam and Pryce,[32] the g tensors and the metal hyperfine splitting tensors of these copper and silver complexes can be related to the molecular orbitals by the equations[15,33-35]

$$g_{\parallel} = 2.0023 - \frac{8\lambda\beta_1^2\beta_2^2}{\Delta_{\parallel}}\left[1 - \left(\frac{\beta_1'}{\beta_1}\right)S - \frac{1}{2}\left(\frac{\beta_1'\beta_2'}{\beta_1\beta_2}\right)T(n)\right] \quad (7)$$

$$g_{\perp} = 2.0023 - \frac{2\lambda\beta_1^2\varepsilon^2}{\Delta_{\perp}}\left[1 - \left(\frac{\beta_1'}{\beta_1}\right)S - \frac{1}{2}\left(\frac{\beta_1'\varepsilon'}{\beta_1\varepsilon}\right)T(n)\right] \quad (8)$$

$$A = P\{-\tfrac{4}{7}\beta_1^2 - \kappa - 2\lambda\beta_1^2[(4\beta_2^2/\Delta_{\parallel}) + \tfrac{3}{7}(\varepsilon^2/\Delta_{\perp})]\} \quad (9)$$

$$B = P[\tfrac{2}{7}\beta_1^2 - \kappa - (22/14)(\lambda\beta_1^2\varepsilon^2/\Delta_{\perp})] \quad (10)$$

where λ is the spin-orbit coupling constant of the free metal ion, κ the Fermi contact term, $\Delta_{\parallel} = E(B_{1g}) - E(B_{2g})$, $\Delta_{\perp} = E(B_{1g}) - E(E_g)$, and P is defined by the equation

$$P = 2\beta\beta_N g_N\langle d_{x^2-y^2}|1/r^3|d_{x^2-y^2}\rangle \quad (11)$$

where g_N is the magnetic moment of the metal isotope and the other symbols denote the usual constants. The only overlap that is considered, because of its great significance, is that between the $d_{x^2-y^2}$ orbital and the normalized nitrogen σ orbitals:

$$S = 2\langle d_{x^2-y^2}|\sigma\rangle \quad (12)$$

In Eq. (12) the ligand σ-function is described as the hybridized function of $2s$ and $2p$,

$$\sigma = \gamma S + (1 - \gamma^2)^{1/2}P_{\sigma} \quad (13)$$

By a suitable treatment of superhyperfine interaction, we can prove that γ is equal to $1/\sqrt{3}$ and that the nature of hybridization of the nitrogen σ orbital is sp^2 (see the next section for details). Also, the ligand coefficient (β_1') in the b_{1g} molecular orbitals of CuTPP and AgTPP has been calculated to be 0.548 and 0.658, respectively (see below). For the calculation of the overlap integral we need suitable wave functions, along with the metal-nitrogen bond distances. In most tetraphenylporphyrins the metal-nitrogen distance has been measured[7] and found to be 2.02 Å. For CuTPP we assume the overlap value calculated by Assour,[15] $S = 0.092$. For the calculation of the $\sigma(b_{1g})$ overlap in AgTPP we used the Ag^{2+} functions of Watson,[36] the nitrogen $2s$ and $2p$ functions of Clementi,[37] and the same metal-nitrogen distance as in CuTPP; S was estimated to be 0.097.

The constant $T(n)$, calculated using the effective nuclear charges on the atom and the metal-ligand internuclear distance, arises from the calculation of the matrix elements of the Hamiltonian with the ligand wave functions. Maki and McGarvey[35] have given the simplified relationship:

$$T(n) = n - [8R(1 - n^2)^{1/2}(z_s - z_p)(z_s z_p)^{5/2}/(z_s + z_p)^5 a_0] \quad (14)$$

where $n = \sqrt{\tfrac{2}{3}}$ for the sp^2 hybrid orbitals of nitrogen, R is the metal-nitrogen internuclear distance, and z_s and z_p are the effective nuclear charges on the nitrogen atom.

The evaluation of $T(n)$ for CuTPP has been made by Assour,[15] who gives $T(n) = 0.33$. The value of $T(n)$ for AgTPP is expected to be quite similar to that in CuTPP. The spin-orbit coupling constants λ for Cu^{2+} and Ag^{2+} have been estimated by Dunn[38] to be -830 and $-1840 \, cm^{-1}$, respectively.

The b_{1g} molecular orbital is normalized, and the normalization condition is

$$(\beta_1)^2 + (\beta_1')^2 - 2\beta_1\beta_1'S = 1 \qquad (15)$$

Since we know the value for β_1' from analysis of the superhyperfine interaction, and S is obtained from the overlap calculation, we are able to calculate β_1, the molecular orbital coefficient of the $d_{x^2-y^2}$ orbital in the b_{1g} antibonding level. We calculated the values $\beta_1 = 0.888$ and 0.820, respectively, for CuTPP and AgTPP. These molecular orbital coefficients provide proof that there is a considerable amount of covalent bonding, particularly for the in-plane σ bonding. All the pyrrole nitrogens, in other words, are very tightly σ-bonded to the metal $d_{x^2-y^2}$ orbital. Moreover, the delocalization of the unpaired electron is considerably larger in AgTPP than in CuTPP. Almost half of the unpaired spin population resides in the ligand orbital. For copper porphyrins Zerner and Gouterman[31] have calculated the electronic population by means of their semiempirical MO calculations. The covalency which they calculated is considerably larger than we have estimated from experimental results. Another thing worth mentioning at this point is that their calculation indicates that the σ-bonding nitrogen orbital is mostly $2p$ with very little $2s$ character. This is in contrast to our findings for these complexes, namely, sp^2 hybridization.

Now we shall make an attempt to calculate the bonding coefficients in the other molecular orbitals. However, for an understanding of these coefficients we need a knowledge of the optical spectra of these complexes, and such studies have been conducted by many authors.[24,39,40] Except for small shifts in energy and intensity, the spectra of the various porphyrins are almost the same. Of interest to our calculations are the spin-allowed transitions $^2B_{1g} \rightarrow {}^2B_{2g}$ and $^2B_{1g} \rightarrow {}^2E_g$. However, for various reasons, there has been considerable difficulty in locating these transitions. Both of the above-mentioned transitions being d–d in character, they will be much weaker in intensity than the extremely intense bands arising out of the ligand π–π transitions. In other words, these ligand transitions arising out of the highly conjugated π system, with extinction coefficients of the order of 10^5, will completely mask the d–d transitions, with low extinction coefficients of the order of 10^2. A look at Zerner and Gouterman's level scheme[31] for the porphyrin complexes will reveal a further complication. Low-energy metal-to-ligand charge-transfer transitions are possible if, as they predict, the ligand $e_g(\pi)$ molecular orbital is directly above the $b_{1g}(\sigma)$ molecular orbital in the copper porphyrin molecules; the same would be true for silver complexes. Also, there is a possibility that there could be ligand-to-metal charge-transfer transitions from low-lying filled ligand levels to the half-filled metal

$b_{1g}(\sigma)$ orbital. The latter two types of transition, arising because of complexing, can be of considerably larger intensity than the pure d–d transitions. For these reasons it can be safely assumed that it would be difficult to pinpoint the transitions we require for our calculations. Moreover, Assour[15] has pointed out the weakness of Thomas and Martell's assignment[39] of the absorption bands near 580 and 560 mμ in copper and nickel derivatives as arising from the d–d transitions. One positive argument against this assignment is the presence of these weak bands even in zinc, palladium, and platinum derivatives. The complete lack of knowledge of these required d–d transitions makes it difficult to calculate the bonding coefficients in the molecular orbitals, other than for the b_{1g} level.

However, it would be possible to compare these bonding coefficients as a function of transition energies, as was done earlier by Roberts and Koski,[10] and later by Assour.[15] Particularly in our case, where we are comparing bonding in CuTPP and AgTPP, this method is much more useful. Here we will calculate the terms $\beta_2^2/\Delta_\parallel$ and $\varepsilon^2/\Delta_\perp$ of Eqs. (7)–(10). Before we describe the results of the calculations an explanation is in order regarding the signs of the metal hyperfine splitting constants. The signs of A and B are negative for ^{63}Cu and ^{65}Cu, but A and B for ^{107,109}Ag turn out to be positive. These are not only derived from our calculations, but are in conformity with earlier results. Also, for the calculation of bonding coefficients we shall use the magnitude of A averaged for the ^{63}Cu and ^{65}Cu, values since we could not resolve the hyperfine components of ^{63}Cu and ^{65}Cu for the B tensor. The ESR spin-Hamiltonian parameters obtained from single-crystal measurements will be used for the calculation of these coefficients.

Using the various values discussed in the foregoing section, and assuming the contributions of the terms associated with $T(n)$ in Eqs. (7) and (8) to be negligible, we have calculated

$$\beta^2/\Delta_\parallel = 0.0380 \times 10^{-3}, \qquad \varepsilon^2/\Delta_\perp = 0.0346 \times 10^{-3} \qquad \text{for CuTPP} \quad (16a)$$

and

$$\beta^2/\Delta_\parallel = 0.0117 \times 10^{-3}, \qquad \varepsilon^2/\Delta_\perp = 0.0153 \times 10^{-3} \qquad \text{for AgTPP} \quad (16b)$$

Substitution of these results into Eqs. (9) and (10) gives the values for P and the Fermi contact term κ. The P values calculated for Cu^{2+} and Ag^{2+} ions in these chelates are, respectively, 350.5×10^{-4} cm^{-1} and -59.8×10^{-4} cm^{-1}, which are in very close agreement with the values computed by McGarvey[41] for these ions. The Fermi contact terms κ have been estimated as 0.355 and 0.69, respectively, for Cu^{2+} and Ag^{2+} ions. Our estimated value for κ in Cu^{2+} seems to be in good agreement with that calculated by Maki and McGarvey,[35] but less than that calculated by Pettersson and Vänngård.[42] However, the value for Ag^{2+} differs considerably from that reported by Pettersson and Vänngård[42] for Ag(II) dithiocarbamates.

The bonding coefficients β_2 and ε in the b_{2g} and e_g molecular orbitals have been studied as a function of the transition energies $\Delta_\parallel$ and $\Delta_\perp$. Figures 10 and 11 show the plots of these coefficients vs the transition energies for

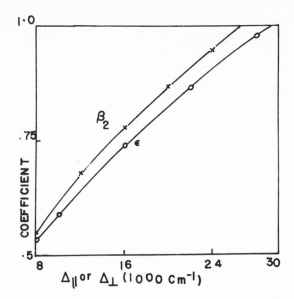

Fig. 10. The molecular orbital coefficients β_2 and ε plotted as functions of the transition energies $\Delta_\parallel$ and $\Delta_\perp$, respectively, for CuTPP.

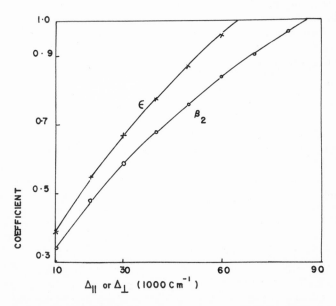

Fig. 11. The molecular orbital coefficients β_2 and ε plotted as functions of the transition energies $\Delta_\parallel$ and $\Delta_\perp$, respectively, for AgTPP.

the CuTPP and AgTPP cases, respectively. Any presence of in-plane and out-of-plane π bonding will give values for β_2 and ε less than 1. If $\beta_2 = 1$ and $\varepsilon = 1$, there is complete absence of in-plane and out-of-plane π bonding. From our calculations, as is revealed by the plots, $\beta_2 = 1$ and $\varepsilon = 1$ will be obtained when $\Delta_\parallel$ and $\Delta_\perp$ are, respectively, 26,300 and 28,900 cm^{-1} for CuTPP. In other words, $\Delta_\parallel < \Delta_\perp$ in the complete absence of in-plane and out-of-plane π bonding. Assour[15] has calculated 22,400 and 17,000 cm^{-1}, respectively, for $\Delta_\parallel$ and $\Delta_\perp$, and according to these values $\Delta_\parallel > \Delta_\perp$; the e_g level is raised above the b_{2g} level. Actually, this difference occurred because of the higher $g_\perp$ value measured by Assour[15] from the polycrystalline sample. Our lower value of $g_\perp$ is more accurate, since it is derived from single-crystal measurements and Q-band measurements of polycrystalline samples; hence our estimate of $\Delta_\parallel < \Delta_\perp$ suggests that the b_{2g} level is slightly above the e_g level, assuming zero in-plane and out-of-plane π bonding. It is quite interesting to note here that in copper acetylacetonate[33] and copper dithiocarbamates[42] such a situation exists, namely $e_g < b_{2g} < a_{1g} < b_{1g}$, although the other situation, $b_{2g} < e_g < a_{1g} < b_{1g}$, exists in etioporphyrin[16] and phthalocyanine complexes.[9] Because of the correct measurement of $g_\perp$ from single-crystal measurements, our estimate of $\Delta_\parallel < \Delta_\perp$ should be correct in the absence of out-of-plane and in-plane π-bonding. However, it is possible for an energy crossover between the b_{2g} and e_g levels to occur in CuTPP. If, with $\Delta_\parallel = \Delta_\perp = 26,300$ cm^{-1} and β_2 still equal to unity, ε becomes equal to 0.95, such a crossover would occur; in other words, a slight out-of-plane π bonding is present, but zero in-plane π bonding. However, we cannot expect the e_g and b_{2g} levels to collapse into one, forming a triply-degenerate level. Here we like to bring into focus the energy-level scheme of Zerner and Gouterman,[31] in which the b_{2g}, e_g, and a_{1g} molecular orbitals in the order of their increasing energy are separated only by about 1000 or 1500 cm^{-1}. Therefore we suggest a value of 24,500 cm^{-1} for $\Delta_\perp$ while $\Delta_\parallel$ remains at 26,300 cm^{-1}. This means $\beta_2 = 1$ and $\varepsilon = 0.92$, implying zero in-plane π bonding, with an inclusion of a small amount of out-of-plane π bonding. This is in contrast to the implication of an energy crossover by Assour on the basis of his ESR results. Moreover, our assumption is in complete agreement with the experimental evidence for vanadyl porphyrin and vanadyl phthalocyanine,[43] in which the unpaired electron is totally localized on the d_{xy} orbital. Also, the semiempirical calculation of Zerner and Gouterman[31] does indicate a small amount ($\varepsilon = 0.92$–0.95) of out-of-plane π bonding in some porphyrin complexes of Mn, Fe, Co, and Ni, while the in-plane π bonding is almost zero ($\beta_2 = 0.99$) for these complexes.

To summarize, the bonding scheme suggested for the CuTPP molecule is $b_{2g} < e_g < a_{1g} < b_{1g}$; very strong in-plane σ bonding with $\beta_1 = 0.88$ and $\beta_1' = 0.548$; zero in-plane π bonding, with $\beta_2 \approx 1$ and $\Delta_\parallel = 26,300$ cm^{-1}; and a small amount of out-of-plane π bonding, with $\varepsilon = 0.92$ and $\Delta_\perp = 24,500$ cm^{-1}. No knowledge of the coefficients in the a_{1g} orbital is possible from the ESR data.

Quite interestingly, the situation is different in AgTPP. On the surface the only difference between CuTPP and AgTPP appears to be the presence of the unpaired electron in the $3d_{x^2-y^2}$ and $4d_{x^2-y^2}$ levels, respectively. However, some change in the whole bonding structure is expected when we go from the $3d^9$ to the $4d^9$ case. This is apparent from our estimate of the values for $\Delta_\parallel$ and $\Delta_\perp$ from the calculated values for $\beta_2^2/\Delta_\parallel$ and $\varepsilon^2/\Delta_\perp$, assuming zero in-plane and out-of-plane π bonding. For AgTPP it turns out that $\Delta_\parallel > \Delta_\perp$ by the above assumption, since the respective estimated transition energies are 85,000 and 65,000 cm^{-1}. However, these values are highly inappropriate for the system under consideration. Although one would expect an increase in these transition energies when we go from the $3d$ to the $4d$ case, the values 85,000 and 65,000 cm^{-1} are definitely too large compared with the transition energies $\Delta_\parallel = 26,300$ and $\Delta_\perp = 24,500$ we had assigned for CuTPP.

It is a well-known fact that the $4d$ ions usually form more covalent bonds than the $3d$ ions. Thus the π bonding, along with a tighter σ bond, would be of greater importance for the $4d$ ions. A tighter σ bond would tend to destabilize the b_{1g} molecular orbital, which would naturally increase the values for $\Delta_\parallel$ and $\Delta_\perp$ in Ag^{2+}. Also, the inclusion of even a small amount of in-plane and out-of-plane π bonding would also tend to destabilize the b_{2g} and e_g orbitals, though slightly less than the earlier case. A net increase in the transition energies $\Delta_\parallel$ and $\Delta_\perp$ will be in order, but it is difficult to decide this net increase without the help of the optical spectra. Unfortunately, no clue as to the transition energies is available from the optical spectrum[24] of AgTPP, which is identical with that of CuTPP, confirming again that all the d–d electronic transitions are masked by the intense transitions of the ring π-system. No suitable system is available for comparison except the dithiocarbamates of copper and silver.[42] It was found by Pettersson and Vänngård[42] that when we go from Cu^{2+} to Ag^{2+} in dithiocarbamate complexes there is a considerable increase in in-plane and out-of-plane π bonding, accompanied by a slight decrease in the metal coefficients of the b_{1g} orbital. These coefficients appear in Table IV. Taking this as a guideline

TABLE IV

Bonding Coefficients in Copper and Silver Dithiocarbamates[a]

	CuDTC	AgDTC
β_1	0.728	0.68
β_2	0.92	0.73
ε	0.99	0.87

[a]From Petersson and Vänngård.[42]

and assuming an increase in AgTPP of about 5000 cm^{-1} from the $\Delta_\parallel$ and $\Delta_\perp$ values in CuTPP, we will obtain $\Delta_\parallel = 31,300$ cm^{-1} and $\Delta_\perp = 29,500$ cm^{-1}. From our estimates of the coefficients β_2 and ε as a function of $\Delta_\parallel$

and $\Delta_\perp$, respectively, we get $\beta_2 = 0.61$ and $\varepsilon = 0.67$. This definitely implies a considerable amount of in-plane and out-of-plane π bonding. Even if the coefficients β_2 and ε in AgTPP, as estimated here, are somewhat in error, it seems certain that there is very strong in-plane and out-of-plane π bonding, because under no circumstances can the $\Delta_\parallel$ and $\Delta_\perp$ values for AgTPP be expected to be larger than our prediction.

We therefore propose the same ordering of the d-energy levels for CuTPP and AgTPP, namely $b_{2g} < e_g < a_{1g} \ll b_{1g}$, with considerable σ-bonding strength. The in-plane π bonding is nil, and the out-of-plane π-bonding is slight in CuTPP, while both are considerably higher in AgTPP; thus in AgTPP $\beta_1 = 0.820$, $\beta_2 \approx 0.61$, and $\varepsilon \approx 0.67$, with $\Delta_\parallel = 31,300\ \mathrm{cm}^{-1}$ and $\Delta_\perp = 29,500\ \mathrm{cm}^{-1}$. The tentative values for the last two parameters need confirmation by a careful low-temperature optical spectral measurement, which might detect the transitions $^2B_{1g} \to {}^2B_{2g}$ and $^2B_{1g} \to {}^2E_g$ as at least small shoulders. Also, spectral studies using polarized light, preferably at low temperatures, would be worthwhile.

SUPERHYPERFINE STRUCTURE

The superhyperfine term in the spin Hamiltonian, Eq. (3), deserves to be examined in detail. A comparison of the ESR spectra of the complexes CuTPP and AgTPP both in solution and in the single crystal indicates that the superhyperfine interaction is mainly isotropic. Superhyperfine interactions in metalloporphyrins and phthalocyanines have been considered in detail by many authors;[15,21,23,33] of these, the treatments of Deal et al.[21] and Abkowitz et al.[23] are a little more specific than the others, so we have used the development of Deal et al.[21] here. In fact, they have assumed cylindrical symmetry with respect to the copper–nitrogen bond, specifying that this assumption would only be valid if the orbital contributions are completely quenched and the unpaired electron is thus a σ electron. However, as they rightly point out, this assumption would only be approximately true if there is a slight unquenching of the orbital momentum by spin-orbit interaction or by other mechanisms.

Using the molecular coordinates of Fig. 7, we can write the superhyperfine part of the spin Hamiltonian, Eq. (3), as

$$\sum_n SA_nI_n = A_N(S_yI_{1y} + S_zI_{1z}) + B_N(S_xI_{1x})$$
$$+ A_N(S_xI_{2x} + S_zI_{2z}) + B_N(S_yI_{2y})$$
$$+ A_N(S_yI_{3y} + S_zI_{3z}) + B_N(S_xI_{3x})$$
$$+ A_N(S_xI_{4x} + S_zI_{4z}) + B_N(S_yI_{4y}) \tag{17}$$

where A_N and B_N are the superhyperfine splitting constants. Here A_N can be measured directly when the static magnetic field is parallel to the z axis of the molecule. In this particular orientation all four nitrogens are equivalent. Similarly, when the field is parallel to the line bisecting the x and y axes in the molecular xy plane all four nitrogens are equivalent and the measured splitting will be equal to B_N.

To produce the σ orbitals directed toward the Cu^{2+} and Ag^{2+} ions, the ligands must have sp^2 hybridized orbitals. We must be able to confirm that this is consistent with our measured superhyperfine splittings. First, we have to make the dipolar corrections to the observed splittings, i.e.,

$$A'_N = A_N - A_D, \qquad B'_N = B_N + 2A_D \qquad (18)$$

where A'_N and B'_N are the values after dipolar corrections, and

$$A_D = g\beta g_N \beta_N / R^3 = 0.2 \times 10^{-4} \quad cm^{-1} \qquad (19)$$

A bond distance of 2.0 Å for R has been assumed. The values obtained in this way may now be used to determine the fractions of s and p character, usually called f_s and f_σ. We may write

$$A'_N = \tfrac{16}{3}\pi\beta g_N \beta_N \phi^2(2s) f_s - \tfrac{4}{5}\beta g_N \beta_N \langle 1/r^3 \rangle_{2p} f_\sigma \qquad (20)$$

$$B'_N = \tfrac{16}{3}\pi\beta g_N \beta_N \phi^2(2s) f_s + 2\tfrac{4}{5}\beta g_N \beta_N \langle 1/r^3 \rangle_{2p} f_\sigma \qquad (21)$$

Using the values of Hurd and Coodin,[44] 5.6 and 3.6 a.u., respectively, for $\phi^2(2s)$ and $\langle 1/r^3 \rangle_{2p}$, we calculate $f_s = 0.025$ and $f_\sigma = 0.46$ for CuTPP and $f_s = 0.036$ and $f_\sigma = 0.075$ for AgTPP. The ratio of f_s to f_σ is consistent with the presence of sp^2 hybridization in the bonding nitrogen orbitals of both CuTPP and AgTPP.

Since we have established the sp^2 hybridization in the bonding nitrogen orbitals, we can calculate the ligand coefficient β'_1 in the b_{1g} molecular orbital given in Eq. (6) from the following expressions:

$$A'_N = (\tfrac{1}{2}\beta'_1)^2 [(16\pi/3)\beta g_N \beta_N \phi^2(2s)(\gamma^2) - \tfrac{4}{5}\beta g_N \beta_N \langle 1/r^3 \rangle_{2p}(1 - \gamma^2)] \qquad (22)$$

$$B'_N = (\tfrac{1}{2}\beta'_1)^2 [(16\pi/3)\beta g_N \beta_N \phi^2(2s)(\gamma^2) + 2\tfrac{4}{5}\beta g_N \beta_N \langle 1/r^3 \rangle_{2p}(1 - \gamma^2)] \qquad (23)$$

where γ is defined by Eq. (13). For sp^2 hybridization we know $\gamma^2 = \tfrac{1}{3}$ and $(1 - \gamma^2) = \tfrac{2}{3}$. The calculated coefficients β'_1 are 0.548 and 0.658, respectively, for CuTPP and AgTPP, indicating a considerable amount of ligand character in the b_{1g} molecular orbital.

A word about the polycrystalline spectra is in order here. The parallel parts of the spectra in polycrystalline CuTPP and AgTPP lead to results identical with those obtained from single-crystal measurements, within experimental error. The hyperfine splitting arrived at by the analysis of this portion is A_N. Interestingly enough, it seems that the perpendicular portions of the polycrystalline spectra are also identical to the values we observe in single crystal at $\theta = 90°$ and $\phi = 45°$. In other words, in CuTPP these spectra consist of 15 equally-spaced lines and in AgTPP they consist of ten equally-spaced lines. This indicates that the perpendicular part of the polycrystalline spectra is a result of interaction with four equivalent nitrogens.

ACKNOWLEDGMENT

This work was supported through a contract with the Army Research Office—Durham. We are indebted to Professor A. Tulinsky for helpful discussion.

REFERENCES

1. A. B. P. Lever, in: *Advances in Inorganic Chemistry and Radiochemistry*, Vol. 7, H. J. Emeleus and A. G. Sharpe, eds. (Academic Press, New York, 1965), p. 28.
2. J. E. Falk, *Porphyrins and Metalloporphyrins* (Elsevier Publishing Co., New York, 1964).
3. D. J. E. Ingram, J. E. Bennett, P. George, and J. M. Goldstein, *J. Am. Chem. Soc.* **78**, 3545 (1956).
4. E. B. Fleischer, *J. Am. Chem. Soc.* **85**, 1353 (1963); **85**, 146 (1963).
5. J. L. Hoard, M. J. Hamor, and T. A. Hamor, *J. Am. Chem. Soc.* **85**, 2334 (1963); **86**, 1938 (1964).
6. S. J. Silvers and A. Tulinsky, *J. Am. Chem. Soc.* **86**, 927 (1964); **89**, 3331 (1967).
7. E. B. Fleischer, C. K. Miller, and L. E. Webb, *J. Am. Chem. Soc.* **86**, 2342 (1964).
8. G. Donnay and C. B. Storm, *Molecular Crystals* **2**, 287 (1967).
9. J. S. Griffith, *Discussions Faraday Soc.* **26**, 81 (1958).
10. E. M. Roberts and W. S. Koski, *J. Am. Chem. Soc.* **82**, 3006 (1959).
11. F. K. Kneübuhl, W. S. Koski, and W. S. Caughey, *J. Am. Chem. Soc.* **83**, 1607 (1961).
12. E. M. Roberts, W. S. Koski, and W. S. Caughey, *J. Chem. Phys.* **34**, 591 (1961).
13. D. Kivelson and S.-K. Lee, *J. Chem. Phys.* **41**, 1896 (1964).
14. A. MacGraph, C. B. Storm, and W. S. Koski, *J. Am. Chem. Soc.* **87**, 1470 (1965).
15. J. M. Assour, *J. Chem. Phys.* **43**, 2477 (1965).
16. E. M. Roberts and W. S. Koski, *J. Am. Chem. Soc.* **83**, 1865 (1961).
17. J. M. Assour and S. E. Harrison, *Phys. Rev.* **136**, A1368 (1964).
18. R. Neiman and D. Kivelson, *J. Chem. Phys.* **35**, 156, 162 (1961).
19. D. J. E. Ingram and J. E. Bennett, *Nature* **175**, 130 (1955); *Discussions Faraday Soc.* **26**, 72 (1958).
20. J. F. Gibson, D. J. E. Ingram, and D. Schonland, *Discussions Faraday Soc.* **26**, 72 (1958).
21. R. M. Deal, D. J. E. Ingram, and R. Srinivasan, in: *Electronic Magnetic Resonance and Solid Dielectrics*, R. Servant and A. Charru, eds. (North-Holland Publishing Co., Amsterdam, 1964), p. 239.
22. J. M. Assour and W. K. Kahn, *J. Am. Chem. Soc.* **87**, 207 (1965).
23. M. Abkowitz, I. Chen, and J. H. Sharp, *J. Chem. Phys.* **48**, 4561 (1968).
24. G. D. Dorough, J. R. Miller, and F. M. Huennekens, *J. Am. Chem. Soc.* **73**, 4315 (1951).
25. C. Rimington, S. F. Mason, and O. Kennard, *Spectrochim. Acta* **12**, 65 (1958).
26. J. M. Goldstein, Ph.D. Dissertation, University of Pennsylvania, 1959.
27. A. Tulinsky, private communication.
28. R. N. Rogers and G. E. Pake, *J. Chem. Phys.* **33**, 1107 (1960).
29. B. Bleaney, *Phil Mag.* **42**, 441 (1951).
30. L. D. Kispert, Thesis, Michigan State University, 1966.
31. M. Zerner and M. Gouterman, *Theoret. Chim. Acta* **4**, 44 (1966).
32. A. Abragam and M. H. L. Pryce, *Proc. Roy. Soc.* (*London*) **A206**, 164 (1951); **A205**, 135 (1951).
33. D. Kivelson and R. Neiman, *J. Chem. Phys.* **35**, 149 (1961).
34. H. R. Gersman and J. D. Swalen, *J. Chem. Phys.* **36**, 3221 (1962).
35. A. H. Maki and B. R. McGarvey, *J. Chem. Phys.* **29**, 31, 35 (1958).
36. R. D. Watson, private communication.
37. E. Clementi, *IBM Research Reports*, R256 (1963).
38. T. M. Dunn, *Trans. Faraday Soc.* **57**, 1441 (1961).
39. D. W. Thomas and A. E. Martell, *Arch. Biochem. Biophys.* **76**, 286 (1958).
40. M. Gouterman, *J. Chem. Phys.* **30**, 1139 (1959).

41. B. R. McGarvey, *J. Phys. Chem.* **71**, 51 (1967).
42. R. Pettersson and T. Vänngård, *Arkiv Kemi* **17**, 249 (1960).
43. J. M. Assour, J. Goldmacher and S. E. Harrison, *J. Chem. Phys.* **43**, 159 (1965).
44. C. M. Hurd and P. G. Coodin, *J. Phys. Chem. Solids* **28**, 523 (1966).

Quadrupole Effects in Electron Paramagnetic Resonance Spectra of Polycrystalline Copper and Cobalt Complexes

Louis D. Rollmann and Sunney I. Chan

Arthur Amos Noyes Laboratory of Chemical Physics*
California Institute of Technology, Pasadena, California

A second-order theory, including quadrupole effects and transitions with $\Delta m_I > 0$, for the calculation of EPR spectra of polycrystalline samples of $S = \frac{1}{2}$ transition-metal ion complexes possessing axial symmetry has been developed. With this theory the EPR spectrum of Cu(acac)$_2$ in Pd(acac)$_2$ and that of Na$_4$CoPTS in DMSO could be explained in detail. The effects of electric quadrupole interaction together with a large anisotropy in the nuclear hyperfine coupling are shown. The advantages of obtaining spectra at two frequencies for the determination of reliable spin-Hamiltonian parameters are demonstrated.

INTRODUCTION

Electron spin resonance spectra of single crystals provide considerable information about the energy-level ordering and the bonding in transition-metal complexes,[1] but the requirement of single crystals places numerous implicit restrictions on the systems which can be studied. As a result, much effort has been devoted to analyzing the spectra of polycrystalline or frozen glass samples, particularly by computer synthesis in recent years.[2,3] All of these calculations, however, have neglected nuclear electric quadrupole effects and transitions involving $\Delta m_I > 0$, even when single-crystal studies have shown that electric quadrupole effects are not negligible relative to the nuclear hyperfine interaction.

Frequently, as has been shown for copper phthalocyanine, absorption peaks may be found in the EPR spectra of polycrystalline samples, which might not have been expected.[4] Such "extra" absorptions are often observed in the spectra of Cu(II) and vanadyl complexes in frozen solution[5] and have been shown clearly in computer-generated spectra of copper complexes to result from large anisotropy in the nuclear hyperfine interaction.[6] The same calculations failed, however, to explain the spectrum of cobalt phthalocyanine (CoPc), and a rhombic distortion or possible quadrupole effect was postulated.

*Contribution No. 3409.

In this chapter the importance of the electric quadrupole interaction on the EPR spectra of frozen glass and powder samples is investigated. Particular attention is directed toward elucidating the contributions of the $\Delta m_I > 0$ transitions and the effects of the quadrupole interaction on the $\Delta m_I = 0$ transitions, particularly in the presence of a large anisotropy in the nuclear hyperfine interaction. The unexplained lines in the reported spectrum of copper acetylacetonate. $Cu(acac)_2$,[5] and the unexplained spectra of cobalt phthalocyanines[6–8] made these systems the logical starting point for the present study. Powder samples of $Cu(acac)_2$ in the palladium analog, $Pd(acac)_2$, were chosen, since single-crystal data for this system have been reported.[1] Frozen solutions of the tetrasodium salt of tetrasulfonated cobalt phthalocyanine (Na_4CoPTS) in dimethylsulfoxide (DMSO) were selected because of its solubility and the absence of intermolecular association in this solvent.[7] The results of this study are felt to be meaningful in view of the potential of EPR for studying transition-metal ions in their biological environment, for investigating unusual oxidation states of these metal ions in inorganic systems, and for elucidating "trapped" intermediates in reactions involving transition-metal complexes— all inaccessible to single-crystal techniques.

LINE SHAPES FOR RANDOM ORIENTATION

For axially symmetrical complexes of transition-metal ions with electron spin $S = \frac{1}{2}$ and nuclear spin $I > \frac{1}{2}$ the spin Hamiltonian is[9]

$$\mathscr{H} = \beta[g_{\parallel}H_zS_z + g_{\perp}(H_xS_x + H_yS_y)] + AS_zI_z + B(S_xI_x + S_yI_y)$$
$$+ Q'[I_z^2 - \tfrac{1}{3}I(I + 1)] - g_N\beta_N H \cdot I \tag{1}$$

where β and β_N are the Bohr and nuclear magneton, respectively; g_N the nuclear g value; $g_{\parallel}$ and $g_{\perp}$ denote the electron g values parallel and perpendicular to the symmetry axis (z axis) of the molecule, respectively; $H_{x,y,z}$ are the components of the magnetic field vector; and S and I denote, respectively, the electron and nuclear spin angular momentum operators. The nuclear hyperfine coupling constants parallel and perpendicular to the symmetry axis are given by A and B, respectively, and Q' is the nuclear-electric quadrupole coupling constant.

The angular dependence of the resonance field for an oriented molecule with axial symmetry has been treated in the paper of Bleaney.[10] This classic paper has formed the basis for the interpretation of single-crystal spectra. From the spin Hamiltonian in Eq. (1) the following expression for the resonance field H may be obtained for the $\Delta m_I = 0$ transitions:

$$H = \frac{h\nu}{g\beta} - \frac{K}{g\beta}m_I - \text{higher-order terms in } A, B, \text{ and } Q' \tag{2}$$

where

$$g^2 = g_{\parallel}^2 \cos^2\theta + g_{\perp}^2 \sin^2\theta \tag{3}$$

$$K^2g^2 = A^2g_{\parallel}^2 \cos^2\theta + B^2g_{\perp}^2 \sin^2\theta \tag{4}$$

and θ is the angle between the molecular symmetry axis and the direction of the applied magnetic field.

In powders or frozen glasses the molecular axes are randomly oriented, the fraction between the angles θ and $\theta + d\theta$ being $\frac{1}{2}\sin\theta\,d\theta$.[11-13] The number of absorbing molecules will thus vary with the angle θ (and hence with the field H), reaching a maximum at $\theta = 90°$ ($H = H_\perp$). If the slight angular dependence of the transition probability is neglected,[14] the intensity of absorption will be proportional to the number of absorbing molecules N, and the rate of change of N with H can be calculated from the differential equation[4]

$$\frac{dN}{dH} = \frac{dN}{d\theta}\frac{d\theta}{dH} = \frac{N_0}{2}\sin\theta\frac{d\theta}{dH} \tag{5}$$

where N_0 is the total number of molecules. The quantity $d\theta/dH$ may be obtained from Eqs. (2)–(4). Since dN is the number of molecules absorbing energy at fields between H and $H + dH$, dN/dH represents the absorption intensity at a given field H. In the absence of quadrupole effects and second-order hyperfine interaction it can be shown from Eqs. (2)–(5) that the absorption intensity is roughly given by

$$\frac{dN}{dH} = \frac{N_0}{2}\frac{g^2}{\cos\theta}\left\{(g_{||}^2 - g_\perp^2)H^0 + \frac{m_I}{\beta}\left[\frac{A^2g_{||}^2 - B^2g_\perp^2}{Kg} - \frac{2K(g_{||}^2 - g_\perp^2)}{g}\right]\right\}^{-1} \tag{6}$$

where $H^0 = h\nu/g\beta$. A similar equation has been derived by Neiman and Kivelson.[4] Subsequent work has extended their treatment to include Lorentzian or Gaussian line shapes[15] and rhombic distortions.[16] In the following sections we generalize Eq. (6) to include quadrupole effects and effects second order in the nuclear hyperfine interaction.

Transitions with $\Delta m_I = 0$. From the work of Bleaney the resonance field for the $\Delta m_I = 0$ transitions, including quadrupole effects and second-order hyperfine interaction, can be shown to be

$$H = H^0 - Km_I - \frac{B^2g_\perp^2}{4H^0g^2}\left[\frac{A^2g_{||}^2 + K^2g^2}{K^2g^2}\right][I(I+1) - m_I^2]$$

$$- \frac{1}{2H^0}\left[\frac{A^2g_{||}^2 - B^2g_\perp^2}{Kg^2}\right]^2\left[\frac{g_{||}g_\perp}{g^2}\right]^2\sin^2\theta\cos^2\theta\, m_I^2$$

$$+ \frac{2Q'^2g_\perp^2\cos^2\theta\sin^2\theta}{Kg^2}\left[\frac{ABg_{||}^2g_\perp^2}{K^2g^4}\right]^2 m_I[4I(I+1) - 8m_I^2 - 1]$$

$$- \frac{Q'^2g_\perp^2\sin^4\theta}{2Kg^2}\left[\frac{Bg_\perp^2}{Kg^2}\right]^4 m_I[2I(I+1) - 2m_I^2 - 1] \tag{7}$$

where now $K^2g^4 = A^2g_{||}^4\cos^2\theta + B^2g_\perp^4\sin^2\theta$, and all coupling constants are expressed in gauss $[A \text{ (gauss)} = A \text{ (ergs)}/g_{||}\beta, B \text{ (gauss)} = B \text{ (ergs)}/g_\perp\beta, K \text{ (gauss)} = K \text{ (ergs)}/g\beta, \text{ and } Q' \text{ (gauss)} = Q' \text{ (ergs)}/g_\perp\beta]$. We note that in Bleaney's original paper, from which the above expression is derived, the perturbation energy denominators in the terms second order in the hyperfine interaction were taken to be $g\beta H^0$, the energy spacing between the successive zeroth-order Zeeman levels at the mean resonance field of the EPR transitions. This assumption is not strictly correct, since the perturbation calculations are rigorously done at the actual resonance fields of the individual transitions, so that the perturbation energy denominators should correspond to the energy spacings between the successive zeroth-order Zeeman levels at the appropriate resonance fields. If this is done, then H^0 in the energy denominators of the second-order hyperfine terms in Eq. (7) would be replaced by H, and the resonance field of a particular transition can be obtained by solving a quadratic equation. We shall ignore this complication here, since the errors inherent in this approximation were found to be quite small for both of the systems considered in this work. The terms second order in the hyperfine interaction cannot be neglected, since these terms and the quadrupole terms are generally comparable in magnitude. We note, however, that the quadrupole terms are independent of the magnetic field.

We now consider Eq. (7) in two important limits. At $\theta = 0°$ (parallel orientation) $g \to g_\|$, $K \to A$, and only the first three terms contribute, the third becoming $(B^2 g_\perp^2 / 2H_\|^0 g_\|^2)[I(I + 1) - m_I^2]$. Hence the quadrupole terms do not contribute to the positions of the resonance fields in the parallel orientation. In the perpendicular limit, where $g \to g_\perp$ and $K \to B$, the second-order hyperfine terms reduce to $[(A^2 g_\|^2 + B^2 g_\perp^2)/4H_\perp^0 g_\perp^2][I(I + 1) - m_I^2]$, and it is evident that second-order hyperfine interaction becomes more important at $\theta = 90°$ than at $\theta = 0°$, if the hyperfine interaction is highly anisotropic such that $A \gg B$. When $\theta = 90°$ the last term in Eq. (7) must also be included, and this quadrupole contribution to the resonance field simplifies to $(Q'^2/2B)m_I[2I(I + 1) - 2m_I^2 - 1]$. We see that the importance of the quadrupole contribution at $\theta = 90°$ depends upon the ratio Q'/B. The remaining terms in Eq. (7) contribute to the resonance field positions only at intermediate angles, orientations which do not normally make significant contributions to the spectrum of polycrystalline or powder samples, except where angular anomalies result in "extra" absorptions. When $A \gg B$ the effect of these terms on the field positions of the "extra" absorptions can be appreciable, particularly when the angular anomalies occur near 70° or 80°.

Differentiating Eq. (7), we obtain the following expression for dN/dH:

$$\frac{dN}{dH} = \frac{N_0}{2}\left(\frac{g^2}{\cos\theta}\right)\left\{(g_\|^2 - g_\perp^2)H^0 + m_I\left[\frac{(A^2 g_\|^4 - B^2 g_\perp^4)}{Kg^2} + 2K(g_\perp^2 - g_\|^2)\right]\right.$$

$$- [I(I + 1) - m_I^2]\frac{A^2 B^2 g_\|^2 g_\perp^2}{4H^0 g^4}\left[\frac{2(A^2 g_\|^4 - B^2 g_\perp^4)}{K^4 g^2} + \frac{(g_\perp^2 - g_\|^2)}{K^2}\right]$$

$$+ [I(I + 1) - m_I^2]\frac{B^2 g_\perp^2}{4H^0 g^2}(g_\perp^2 - g_\|^2)$$

$$- \frac{(A^2 g_\|^2 - B^2 g_\perp^2)^2 g_\|^2 g_\perp^2 m_I^2}{H^0 g^4}\frac{(\cos^2\theta - \sin^2\theta)}{K^2 g^2}$$

$$-\frac{(A^2g_\parallel^2 - B^2g_\perp^2)^2 g_\parallel^2 g_\perp^2 m_I^2}{2H^0 g^4}\left[\frac{2(A^2g_\parallel^4 - B^2g_\perp^4)}{K^4 g^6} - \frac{3(g_\perp^2 - g_\parallel^2)}{K^2 g^4}\right]\sin^2\theta\cos^2\theta$$

$$+\frac{2Q'^2 A^2 B^2 g_\parallel^4 g_\perp^6 m_I}{g^6}[4I(I+1) - 8m_I^2 - 1]$$

$$\times\left[\frac{2(\cos^2\theta - \sin^2\theta)}{K^5 g^2} + \frac{5(A^2 g_\parallel^4 - B^2 g_\perp^4)}{K^7 g^6}\sin^2\theta\cos^2\theta\right]$$

$$-\frac{Q'^2 B^4 g_\perp^{10} m_I}{2g^6}[2I(I+1) - 2m_I^2 - 1]$$

$$\times\left[\frac{4\sin^2\theta}{K^5 g^2} + \frac{5(A^2 g_\parallel^4 - B^2 g_\perp^4)}{K^7 g^6}\sin^4\theta\right]\Bigg\}^{-1} \tag{8}$$

This equation, although lengthy, can be readily programmed on a digital computer. The angular dependence of the transition probability can be included in the calculations by multiplying Eq. (8) by $g_\perp^2[(g_\parallel^2/g^2) + 1]$ for the $\Delta m_I = 0$ transitions.[14]

Transitions with $\Delta m_I = \pm 1$, ± 2. As discussed by Bleaney,[10] the $\Delta m_I = \pm 1$, ± 2 transitions may have considerable intensity. It is from these transitions, as a result of the direct interaction of the nucleus with the magnetic field, that the relative signs of A, B, and Q' may be obtained, since the resonance field for $\Delta m_I = 0$ transitions is independent of the sign of A, B, and Q'. The nuclear Zeeman term, though small, cannot be neglected, since it can result in shifts of the order of several gauss in the resonance fields of these transitions at magnetic fields of 10,000 G. The resonance fields for the $\Delta m_I > 0$ transitions can be calculated from the expressions given by Bleaney. However, in adapting his results we have evaluated the nuclear Zeeman term at the average resonance field H^0 of the EPR transitions rather than at the actual resonance field of each transition, H. This approximation results in considerable simplification in the evaluation of dN/dH. The errors inherent in this approximation are small, since the resonance fields for these transitions do not generally differ from H^0 by more than 10–20%, particularly at higher magnetic fields, where the nuclear Zeeman energy becomes more important.

Transitions with $\Delta m_I = \pm 1$. In discussing $\Delta m_I = \pm 1$ transitions we adopt the notation of Bleaney[10] by substituting k for m_I in Eq. (7), where $k = (I - \frac{1}{2})$, $(I - \frac{3}{2}),\ldots, -(I - \frac{1}{2})$. The resonance field for transitions $(m_s, k \pm \frac{1}{2}) \to (m_s - 1, k \mp \frac{1}{2})$ can be found by adding to Eq. (7) the quantity

$$\pm\left\{\frac{kQ'g_\perp}{g}\left[\frac{3A^2 g_\parallel^4}{K^2 g^4}\cos^2\theta - 1\right] - \frac{g_N\beta_N H^0}{g\beta}\frac{(Ag_\parallel^2\cos^2\theta + Bg_\perp^2\sin^2\theta)}{Kg^2}\right\} \tag{9}$$

where H^0 has been substituted for H in Bleaney's original expression, as described above. The final expression for the resonance field can again be differentiated. The extra terms from Eq. (9) yield the following additional term in dN/dH:

$$\mp \frac{N_0}{2}\left(\frac{g^2}{\cos\theta}\right)\left\{\frac{3kQ'A^2g_\parallel^4g_\perp}{g^3}\left[\frac{2}{K^2} - \frac{2(A^2g_\parallel^4 - B^2g_\perp^4)}{K^4g^4}\cos^2\theta\right.\right.$$

$$+ \frac{(g_\perp^2 - g_\parallel^2)}{K^2g^2}\cos^2\theta\Bigg]$$

$$- \frac{kQ'g_\perp}{g}(g_\perp^2 - g_\parallel^2) + \frac{g_N\beta_N H^0}{K\beta}\left[\frac{-2(Ag_\parallel^2 - Bg_\perp^2)}{g}\right.$$

$$+ \frac{(Ag_\parallel^2\cos^2\theta + Bg_\perp^2\sin^2\theta)}{K^2g^5}(A^2g_\parallel^4 - B^2g_\perp^4)$$

$$\left.\left.\left. + \frac{2(Ag_\parallel^2\cos^2\theta + Bg_\perp^2\sin^2\theta)}{g^3}(g_\parallel^2 - g_\perp^2)\right]\right]\right\}^{-1} \qquad (10)$$

where the appropriate sign relative to Eq. (9) is indicated. The absorption intensity is obtained, relative to that of $\Delta m_I = 0$ transitions, by multiplying dN/dH by the factor

$$4k^2\{(I + \tfrac{1}{2})^2 - k^2\}(4Q'^2g_\perp^2/K^2g^2)(ABg_\parallel^2g_\perp^2/K^2g^4)^2\cos^2\theta\sin^2\theta \qquad (11)$$

Because of the $\cos^2\theta\sin^2\theta$ factor, these transitions have zero intensity at both $\theta = 0°$ and $\theta = 90°$. At intermediate angles the absorption intensity also depends on the angular dependences of K and g. In practice, the angular variations in K are greater than the corresponding variations in g, and hence influence the absorption intensities to a significantly greater extent. If the hyperfine interaction is isotropic, the intensity factor is peaked near $\theta = 45°$. When $A \gg B$ the region of maximum absorption intensity, however, will generally be found at $\theta > 70°$. Since the intensity factor vanishes for $k = 0$, the $(m_s, \pm\tfrac{1}{2}) \to (m_s - 1, \mp\tfrac{1}{2})$ transitions can be ignored.

Transitions with $\Delta m_I = \pm 2$. For the transitions $(m_s, m_I \pm 1) \to (m_s - 1, m_I \mp 1)$, where m_I can have values $(I - 1), (I - 2),\ldots, -(I - 1)$, the resonance field is given by Eq. (7) plus the following additional term:

$$\pm\left\{\frac{2m_IQ'g_\perp}{g}\left[\frac{3A^3g_\parallel^4}{K^2g^4}\cos^2\theta - 1\right] - \frac{2g_N\beta_N H^0}{g\beta}\frac{(Ag_\parallel^2\cos^2\theta + Bg_\perp^2\sin^2\theta)}{Kg^2}\right\} \qquad (12)$$

As in the $\Delta m_I = \pm 1$ transitions, the nuclear Zeeman interaction is calculated at H^0 instead of at the actual resonance field of each transition. Since Eqs. (9) and (12) differ formally by only a factor of two, the additional term in dN/dH for $\Delta m_I = \pm 2$ transitions may be obtained by substituting m_I for k in Eq. (10) and multiplying by two. For the $\Delta m_I = \pm 2$ transitions, however, the intensity factor multiplying the resulting equation for dN/dH will be

$$\{(I + 1)^2 - m_I^2\}\{I^2 - m_I^2\}(Q'^2g_\perp^2/4K^2g^2)(Bg_\perp^2/Kg^2)^4\sin^4\theta \qquad (13)$$

In contrast to the intensities of the $\Delta m_I = \pm 1$ transitions, which go to zero at $\theta = 90°$, the $\Delta m_I = \pm 2$ transitions reach their maximum intensity at this angle.

It is evident from the intensity factors, Eqs. (11) and (13), that the $\Delta m_I > 0$ transitions have zero intensity in the absence of quadrupole interaction. For a given Q' the intensity factor varies as $(Q'/B)^2(A/B)$ for $\Delta m_I = \pm 1$, and as $(Q'/B)^2$ for $\Delta m_I = \pm 2$, near the perpendicular limit where the contributions of these transitions are generally found to be most important in polycrystalline spectra. Although the ratio Q'/B may be small for a small quadrupole interaction, the intensity factors may not be completely negligible when I is large and when $A \gg B$. Furthermore, the absorption intensity of the $\Delta m_I > 0$ transitions may assume surprisingly large values as a result of unusually large $d\theta/dH$'s.

EXPERIMENTAL

Materials

The preparation, purification, and analysis of Na_4CoPTS have been described previously.[7] DMSO (Baker, reagent) was used without further purification, Karl Fischer titration showing the water content to be about $10^{-2}\ M$. No changes in the results were observed when DMSO, dried by fractional distillation from CaH_2 under vacuum, was used. Solutions studied ranged from 0.008 to 0.08 M Na_4CoPTS.

$Cu(acac)_2$ and $Pd(acac)_2$ were prepared by the standard methods,[17] purified by multiple recrystallization from chloroform, and dried at 56° *in vacuo* over P_2O_5 to constant weight. A second sample of $Cu(acac)_2$ was purified by multiple recrystallization from 50% ethanol. Although the preparative methods are well established, the copper content was checked electrolytically and found to be 24.19% (calc. 24.28%). Samples were prepared by dissolving weighed amounts of $Cu(acac)_2$ and $Pd(acac)_2$ in chloroform, filtering (as a precaution), and evaporating the solvent under a nitrogen stream, again drying the product powders *in vacuo* over P_2O_5. Concentrations ranged from 0.34 to 1.9 mole % $Cu(acac)_2$ in $Pd(acac)_2$ (0.29 to 1.6% by weight).

Instrumentation

ESR data were obtained on a Varian Model V-4500 spectrometer at X-band (9.525 kHz) and K-band (34.77 kHz) frequencies with 100-kHz field modulation. The magnetic field was determined using an Alpha Model 675 NMR Gaussmeter. The *g* values were obtained both by comparison with the resonance position of a 0.1% dispersion of diphenylpicrylhydrazyl (DPPH) in solid KCl and by calculation from the measured microwave frequency, the two methods agreeing to about 0.03%.

In the studies at X-band quartz sample tubes 3 mm in diameter were employed, and for work near 77°K the sample temperature was regulated by a Varian Model V-4540 variable-temperature controller plus accessories.

At the higher K-band frequency much smaller samples were required, and 1-mm-diameter, thin-walled quartz tubes were used. Here the entire K-band cavity was cooled directly by liquid nitrogen in the investigations at 77°K. In all the low-temperature studies the samples were precooled by immersion in liquid nitrogen prior to transfer to the cavity.

Computer Program

A program based on Eqs. (7)–(13) was written for the IBM 7090/7094 computer to calculate the resonance fields and absorption intensities. All intensities were normalized relative to that of the $\Delta m_I = 0$ transition with $m_I = -I$, $\theta = 0°$. At angles where very large values of dN/dH were obtained, the intensities were estimated from the corresponding values at angles 1° removed. For a given set of $g_{\parallel}$, $g_{\perp}$, A, B, Q', and the resonance field of DPPH the resonance fields and the intensities of all the transitions were calculated for values of θ at 1° or 10° intervals, as appropriate. The starting set of parameters was then adjusted until an apparent best fit was obtained on the basis of a δ-function line shape.

In order to facilitate comparison with the experimentally observed spectra, a theoretical spectrum was generated by a second program, written to account for finite linewidths. Using the spin-Hamiltonian parameters giving the best apparent fit on the basis of a δ-function lineshape, the field position and intensity of each transition were calculated at 1° intervals from $\theta = 0°$ to $\theta = 90°$. A Lorentzian line with area equal to the intensity of each transition at each field position was then generated by the computer program at 1-G intervals using a linewidth estimated from those of the $\Delta m_I = 0$ transitions in the parallel orientation. These individual Lorentzians were then summed to give a resultant absorption spectrum which, in the case of $Cu(acac)_2$, was also differentiated to yield a final derivative spectrum for comparison with experiment.

RESULTS

EPR data for $Cu(acac)_2$ in $Pd(acac)_2$ were obtained at both 298° and 77°K. Frozen solutions of Na_4CoPTS in DMSO were only investigated at 77°K.

Table I lists the EPR results obtained, as described below, for $Cu(acac)_2$ and for Na_4CoPTS and includes a comparison with values reported previously. Where it has been possible to resolve a rhombic component ($g_1 \neq g_2$) this has been indicated. Otherwise, $g_1 = g_2 = g_{\perp}$.

$Cu(acac)_2$

The X-band EPR spectrum of a polycrystalline sample of $Cu(acac)_2$ in $Pd(acac)_2$ at 298°K is presented in Fig. 1. At liquid-nitrogen temperature (77°K) the lines become somewhat narrower, but no further structure is resolved. The changes in the g values and in the nuclear hyperfine coupling constants with changing temperature illustrated in Table I appear to be real,

TABLE I
EPR Data for Copper Acetylacetonate and Cobalt Phthalocyanine[a]

Compound	Solvent	T(°K)	$g_\parallel$	A	g_1	B_1	g_2	B_2	Q'	Ref.
Cu(acac)₂	Pd(acac)₂ [b]	298	2.264	−0.0182	2.051	−0.0021	2.054	−0.0020	+0.0008	(d)
			±0.002	±0.0002	±0.002	±0.0002	±0.002	±0.0002	±0.0001	
	Pd(acac)₂ [c]	77	2.261	−0.0188	2.048	−0.0025	2.052	−0.0024	+0.0008	(e)
	CHCl₃/C₇H₈	300	2.266	−0.0160	2.052	−0.00195	2.055	−0.00195	+0.0007	(f)
	CHCl₃	77	2.264	0.01455	2.036	0.029	—	—	—	(g)
	CHCl₃	77	2.2875	−0.0173	2.051	−0.00227	—	—	—	(h)
	CHCl₃	77	2.285	0.0175	2.042	0.00282	—	—	—	
Na₄CoPTS	DMSO	77	2.005	(+)0.0098	2.257	(−)0.0021	—	—	(−)0.0002	(d)
			±0.002	±0.0002	±0.005	±0.0005			±0.0001	
CoPc	H₂SO₄	77	2.027	0.00849	(2.30)	(0.0012)	—	—	—	(i)
			±0.003	±0.00005	±0.02					
	H₂SO₄	77	2.029	0.0085	2.546	0.0096	—	—	—	(j)
	C₅H₅N	77	2.016	0.0078	2.268	—	—	—	—	(j)
	α-ZnPc[b]	300	2.007	0.0116	2.422	0.0066	—	—	—	(k)
		77	±0.003	±0.0003	±0.003	±0.0003				

[a]All coupling constants in cm^{-1}; estimated values in parentheses. Unless specified, only the absolute values of the coupling constants are given.
[b]Powder sample.
[c]Single crystal.
[d]Present work.
[e]Maki and McGarvey.[1]
[f]Gersmann and Swalen.[5]
[g]R. Wilson and D. Kivelson, J. Chem. Phys. 44, 4445 (1966).
[h]H. A. Kuska, M. T. Rogers, and R. E. Drullinger, J. Phys. Chem. 71, 109 (1967).
[i]Vänngård and Aasa.[6]
[j]Assour.[8]
[k]Assour and Kahn.[25]

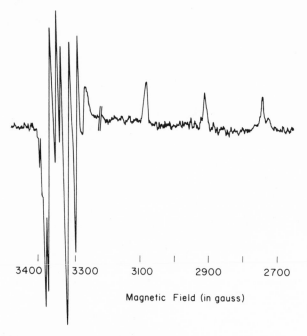

Fig. 1. Derivative X-band EPR spectrum of polycrystalline
Cu(acac)$_2$ in Pd(acac)$_2$ at 298°K. Amplification increased approximately fourfold below 3200 G.

the error limits representing the uncertainty in the absolute values. From
the earlier discussion of line shapes of polycrystalline spectra it is apparent
that the components at higher field represent the envelope of absorption of
molecules whose symmetry axes are nearly perpendicular to the field. The
spectrum is quite similar to that reported by previous workers for Cu(acac)$_2$
in frozen toluene/CHCl$_3$ solution,[5] except for the smaller linewidths. At the
high-field extreme the sharp weak peak near $H = 3400$ G gives the position
of the DPPH standard. The shoulder near this resonance has been attributed
to a ^{65}Cu "extra" absorption resulting from an angular anomaly in the
$m_I = +\frac{3}{2}$ hyperfine component.

   The parallel components are readily observed at lower field in Fig. 1.
When the spectrum was expanded structure due to ^{65}Cu could be resolved
for all the parallel components at concentrations below about 1 mole %
Cu(acac)$_2$. These were symmetrically displaced about the ^{63}Cu components
with $A(^{65}$Cu)/$A(^{63}$Cu) = 1.07, as expected.[18] In Fig. 1 the ^{65}Cu splitting is
clearly resolved for the lowest-field hyperfine component.

   The complex region about $H_\perp^0$ has been expanded and depicted in
Fig. 2. A theoretical line spectrum, calculated using the spin-Hamiltonian
parameters given in Table I assuming a δ-function line shape and with

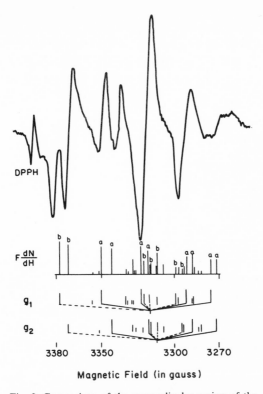

Fig. 2. Comparison of the perpendicular region of the X-band spectrum of polycrystalline Cu(acac)$_2$ in Pd(acac)$_2$ with the theoretical line spectrum calculated using the spin-Hamiltonian parameters given in Table I. To illustrate the rhombic asymmetry, the resonance fields for the two sets of rhombic components (g_1 and g_2) are presented separately. The field positions of the $\Delta m_I = 0$ transitions are connected to the appropriate $H_\perp^0$ by solid or broken lines, with the broken lines designating the "extra" absorptions. The designations a and b have also been used to identify these resonances to facilitate comparison with the calculated spectra presented in Fig. 4. Of the remaining field position markers, the smaller markers identify the $\Delta m_I = \pm 1$ transitions and the larger markers designate the $\Delta m_I = \pm 2$ transitions.

appropriate contributions from all transitions including those involving $\Delta m_I > 0$, is shown for comparison. The calculated intensity of each transition is given by $F(dN/dH)$, where F represents the appropriate multiplicative intensity factor discussed earlier. The ^{65}Cu transitions have not been included in the calculations, since with one exception these splittings are

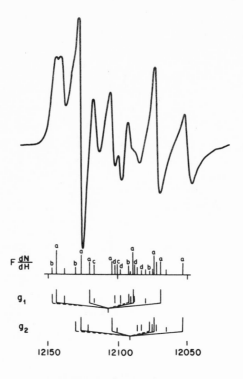

Magnetic Field (in gauss)

Fig. 3. Comparison of the perpendicular region of
the K-band spectrum of polycrystalline Cu(acac)$_2$
in Pd(acac)$_2$ with the theoretical line spectrum
calculated using the spin-Hamiltonian parameters
given in Table I. See caption to Fig. 2 for description of legend. The designations c and d
identify the $\Delta m_I = \pm 1, \pm 2$ transitions, respectively, to facilitate comparison with the calculated
spectra presented in Fig. 5.

not resolved in the perpendicular spectrum. For the small values of B and
Q' given in Table I the calculated separations between the corresponding
^{63}Cu and ^{65}Cu transitions in the perpendicular region are well within the
linewidths of the observed resonances. From earlier work[18] one would
expect $B(^{65}$Cu$)/B(^{63}$Cu$) = 1.07$, and $Q'(^{65}$Cu$)/Q'(^{63}$Cu$) = 1.08$. To illustrate
the rhombic asymmetry, the resonance fields for the two sets of rhombic
components (g_1 and g_2) are also presented separately. The resonance
positions of the $\Delta m_I = 0$ transitions are connected to the appropriate
$H_\perp^0(h\nu/g_\perp\beta)$ by solid or broken lines, with the broken lines designating the
"extra" absorptions.

The K-band spectrum at 298°K in the region about $H_\perp^0$ is shown in Fig. 3. A theoretical line spectrum calculated with the spin-Hamiltonian parameters given in Table I is again included for comparison. As in the case of the theoretical X-band spectrum, no ^{65}Cu transitions are included. To emphasize the increased importance of the rhombic asymmetry at the higher magnetic fields, the two sets of rhombic components are again illustrated separately.

The four parallel components, though not shown in Fig. 3, were clearly observed in the K-band spectrum. At 298°K the ^{65}Cu satellites were resolved in the $m_I = \pm\frac{3}{2}$ hyperfine components and appeared as shoulders in the $m_I = \pm\frac{1}{2}$ components. At 77°K all four ^{65}Cu components were fully resolved. Finally, as in the X-band spectrum, there was no apparent dependence of the line shape of the parallel components on m_I.

In the analysis of the data it was assumed that the field positions at $\theta = 90°$ and those for the extra absorptions correspond roughly to points in the observed spectrum at which the derivative approaches zero. This assumption is probably justified, since in the limit of $\theta = 90°$ and where "angular anomalies" occur the calculated intensities diverge. This assumed correspondence is most evident upon comparison of the calculated field positions with the experimental X-band spectrum in Fig. 2 in the spectral region near about 3380 G for the "extra" absorptions and near 3350 G for the $m_I = \frac{3}{2}$ component of the $\Delta m_I = 0$ transitions. The spectrum at K-band lends support to this correspondence, the differences between the observed and calculated field positions being covered by the error limits indicated in Table I.

In contrast to the above procedure used to locate the resonance fields of the transitions in the perpendicular limit, the spectrum corresponding to the parallel orientation was analyzed, using the field positions of the resonance maxima of the hyperfine components in the derivative presentation. The spin-Hamiltonian parameters extracted in this manner were found to best fit both the X-band and K-band spectra for $\theta = 0°$, provided the same criterion was used to locate the resonance fields of the hyperfine components in both spectra.

In Fig. 4 two calculated spectra are presented and are compared with the observed X-band spectrum in the perpendicular region. In both cases a Lorentzian line-shape function with a 5-G linewidth (full width at half-height) was used in synthesizing the spectrum. The two computed spectra differ merely in the choice of the quadrupole coupling constant Q'. Otherwise, the values of the remaining spin-Hamiltonian parameters indicated in Table I were employed. In one of these calculations $Q' = 0.0008\ \text{cm}^{-1}$, and in the other Q' was set identically equal to zero. The differences between the calculated spectra in the presence and in the absence of quadrupole interaction can be seen to be quite small, and hence it appears that no reliable quadrupole coupling constant can be obtained from the X-band data alone. The "extra" absorptions at 3370–3380 G result from angular anomalies at $\theta = 66°$. Since the quadrupole effects are negligible at this angle, the positions of these

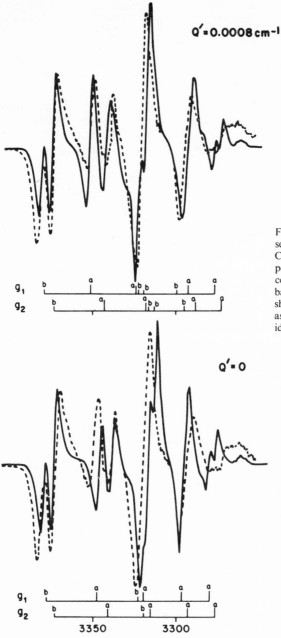

$Q' = 0.0008\ cm^{-1}$

$Q' = 0$

Magnetic Field (in gauss)

Fig. 4. Comparison of the observed X-band spectrum of $Cu(acac)_2$ in $Pd(acac)_2$ in the perpendicular region with two computer-synthesized spectra based on Lorentzian line shapes. See Fig. 2 for detailed assignment of the spectrum and identification of the transitions.

resonances therefore exhibit little dependence on Q'. The inclusion of the quadrupole interaction has the effect of increasing the separation between $\Delta m_I = 0$ hyperfine components bearing the same $|m_I|$. Since $I = \frac{3}{2}$, there is however, no effect on the separation between the adjacent $|m_I| = \frac{3}{2}$ and $|m_I| = \frac{1}{2}$ components. Due to the considerable overlap of the rhombic components at X-band, the effects of quadrupole interaction are only evident in that part of the spectrum in the vicinity of 3270 G ($m_I = -\frac{3}{2}$) and in the region near 3340 G ($m_I = +\frac{3}{2}$). In Fig. 4 the agreement between the observed and calculated spectra in the region near 3270 G can be seen to be considerably improved by the inclusion of a finite Q'. The rather ambiguous fit near 3340 G, however, precludes determination of a reliable quadrupole coupling constant from these X-band data alone.

Calculated K-band spectra with and without the inclusion of quadrupole interaction are compared with the observed spectrum in the region near $H_\perp^0$ in Fig. 5. In contrast to the X-band results, the differences in the two calculated spectra are quite striking. This marked difference can be attributed to two factors. First of all, the frequency of occurrence of "angular anomalies" as well as the field positions of these "extra" absorptions become more sensitive to the quadrupole interaction at K-band. Angular anomalies occur less frequently at the higher frequency, as is evident in Fig. 6, where the angular dependences of the resonance fields at X-band and K-band are compared for the rhombic components centered at g_1. Mathematically, the angular anomalies result from the different angular dependences of g and K, and the resultant simplification of the spectrum at K-band reflects the decreased importance of the nuclear hyperfine interaction relative to the Zeeman interaction. Consequently, angular anomalies at K-band are generally confined to orientations nearer to the perpendicular limit, where the effects of any quadrupole interaction would be more important in determining the field positions. Frequently the existence of an angular anomaly for a particular transition may also depend upon the presence or absence of the quadrupole coupling because of the strong angular dependence of the quadrupole terms near the perpendicular limit. This behavior is clearly manifested in the K-band spectrum near 12,150 G. Two resonances fall in this spectral region: the $m_I = +\frac{3}{2}$ hyperfine component at $\theta = 90°$, and an "extra" absorption resulting from an angular anomaly at $\theta = 87°$ for the same transition. The spectrum calculated with $Q' = 0.0008 \text{ cm}^{-1}$ clearly reproduces the observed splitting between these resonances, whereas for $Q' = 0.0000 \text{ cm}^{-1}$ the "extra" absorption is absent in the computed spectrum. This same behavior was noted for the $m_I = -\frac{1}{2}$ hyperfine component, even though the presence or absence of the angular anomaly here is not particularly evident in the theoretical spectrum. Secondly, because of the greater importance of the rhombic asymmetry at higher magnetic fields, the rhombic components are more widely separated at K-band, facilitating more accurate determination of the resonance field positions of the individual rhombic components. At X-band the rhombic components overlap to an appreciable extent, and it was not possible to locate the resonance field positions of the

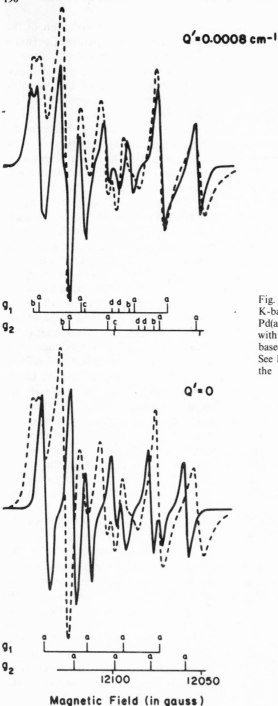

Fig. 5. Comparison of the observed K-band spectrum of Cu(acac)$_2$ in Pd(acac)$_2$ in the perpendicular region with two computer-simulated spectra based on Lorentzian line shapes. See Fig. 3 for detailed assignment of the spectrum and identification of the transitions.

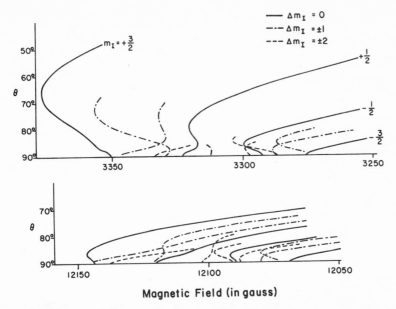

Magnetic Field (in gauss)

Fig. 6. Angular dependence of the resonance fields of the $\Delta m_I = 0, \pm 1, \pm 2$ transitions near the perpendicular limit for the g_1 rhombic components of $Cu(acac)_2$ in $Pd(acac)_2$ at both X-band and K-band.

individual components sufficiently reliably to ascertain the resonance shifts resulting from the quadrupole interaction, even in spectral regions where the resonances are not masked or complicated by the presence of "extra" absorptions or $\Delta m_I > 0$ transitions. This is not the case at K-band. The $m_I = -\frac{3}{2}$ hyperfine component of the g_2 rhombic transitions, e.g., is well separated from the rest of the spectrum, and there can be no ambiguity regarding the fit between the calculated and observed resonance field for this transition. As seen in Fig. 5, the overall agreement between the calculated and observed K-band spectra is excellent when quadrupole interaction is included in the calculations. The successful prediction of the $\Delta m_I > 0$ contributions near 12,100 G is gratifying, and lends further credence to the present theoretical treatment and the interpretation of the spectral results. Thus by a comparison of the EPR spectra obtained at X-band and K-band it has been possible to identify "extra" absorptions, and to distinguish between effects due to rhombic asymmetry and electric quadrupole interaction.

Na_4CoPTS

The X-band spectrum of Na_4CoPTS in DMSO at 77°K is shown in Fig. 7. The region below $H = 2900$ G was carefully searched for the weak lines previously reported by Assour[8] but absent in earlier works.[6] No resonances were observed from 2900 to below 1000 G.

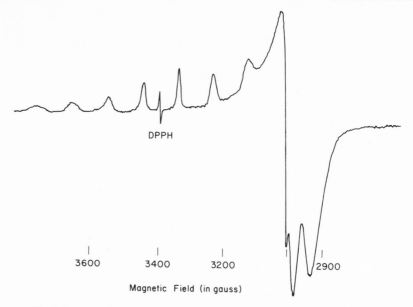

Fig. 7. Derivative X-band spectrum of 0.05 M Na$_4$CoPTS in DMSO at 77°K.

The parallel hyperfine components, centered roughly about the DPPH signal, are clearly discernible, with the exception of the $m_I = +\frac{7}{2}$ component (A taken to be positive), which is buried under the perpendicular transitions. The observed variation in the signal heights and widths of the parallel components is particularly striking. As in the case of Cu(acac)$_2$, $g_{\parallel}$ and A were determined from the $g_{\parallel}$ spectrum using the field positions of the resonance maxima of the hyperfine components in the derivative presentation.

In the perpendicular region the spectrum is not well resolved, though some structure is observed. From the resonance position and the spectral width B and $g_{\perp}$ can both be estimated; however, a more reliable determination of these interaction constants is obtained from a combined analysis of the X-band and K-band data, and from actual computer synthesis of the spectrum. Due to the simplicity of the derivative spectrum about $H_{\perp}^0$, this region of the spectrum may be integrated with some reliability. In Fig. 8 this region, expanded and integrated, is compared with two theoretical absorption spectra calculated using the spin-Hamiltonian parameters given in Table I, except that the quadrupole coupling constant Q' was set identically equal to zero in the calculations of one of these spectra. While the nuclear hyperfine coupling constant A is almost certainly positive[19] we have taken B to be negative.[20] The observed m_I-dependence of the linewidths noted above for the parallel components was assumed in these calculations. The full linewidth at half-height of the $m_I = \pm\frac{1}{2}$, $\Delta m_I = 0$ transitions was taken to be 20 G. The $\Delta m_I > 0$ transitions were assumed to have linewidths comparable to those of the $\Delta m_I = 0$ transitions. Because of the above assumptions regarding the linewidths, it was not possible to exclude, on the basis of these calculations, the possibility of a positive B value. If B were taken to be positive, only the positions and intensities of the $\Delta m_I > 0$ transitions in Fig. 8 would be altered. The calculations clearly predict a dependence of the field position of the absorption maximum on Q'.

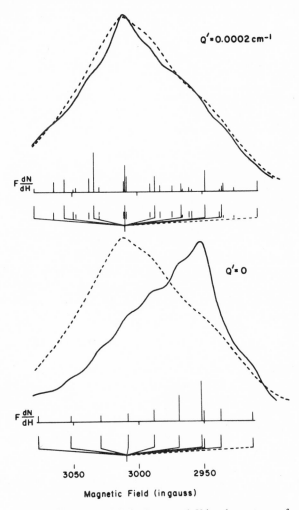

Fig. 8. Comparison of the integrated X-band spectrum of Na$_4$CoPTS in DMSO in the perpendicular region with two computer-synthesized spectra based on Lorentzian line shapes. The calculated line spectra are also shown, with the same field-position markers used to distinguish between the $\Delta m_I = 0, \pm 1, \pm 2$ transitions as in the Cu complex.

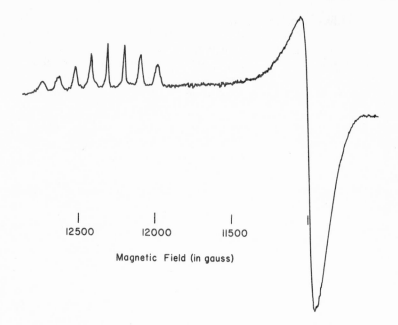

12500	12000	11500

Magnetic Field (in gauss)

Fig. 9. Derivative K-band spectrum of 0.05 M Na$_4$CoPTS in DMSO at 77°K.

As $|Q'|$ is increased the absorption maximum is shifted to higher fields. Thus Q' could be estimated from the field position of the absorption maximum as well as from the shape of the integrated spectrum. This dependence of the absorption maximum and the shape of the absorption spectrum on Q' can be understood largely in terms of the effects of the quadrupole interaction on $d\theta/dH$ (and hence dN/dH) of the $\Delta m_I = 0$ transitions. Because of the large nuclear spin of ^{59}Co, the $\Delta m_I > 0$ transitions can also have sizeable intensities, and these transitions no doubt contribute somewhat to the shape of the spectrum. In the absence of quadrupole interaction the $\Delta m_I > 0$ transitions, of course, have zero intensity.

The K-band spectrum of Na$_4$CoPTS at 77°K is shown in Fig. 9. The variations in the linewidths and signal heights of the parallel components noted in the X-band spectrum are again observed. The data for $\theta = 0°$ confirm the values of A and $g_\parallel$ (Table I) calculated from the X-band spectrum, and this agreement would appear to support our use of the maxima of these parallel components for the determination of these spin-Hamiltonian parameters.

At K-band the parallel and perpendicular regions are well separated. The region about $H_\perp^0$ is very similar to that observed at X-band, but without structure. The peak–peak width of the derivative spectrum (roughly 150 G) suggests a value of about 20 G for B. The derivative spectrum passes through zero at $g = 2.25$. From the position of the absorption maximum and a comparison of the overall width for both the X-band and K-band perpendicular spectra estimates of a possible rhombic distortion from D_{4h} symmetry could be made. The possibility of such a rhombic asymmetry has been included in the larger error limits of $g_\perp$ and B in Table I. On the basis of spectral calculations similar to those presented for the X-band spectrum the effects of the quadrupole interaction on the position of maximum intensity were found to be less pronounced at K-band than at X-band. This result, we believe, arises from the smaller dependences of $d\theta/dH$ on the quadrupole interaction at the higher frequency.

DISCUSSION

The effects of second-order hyperfine interaction and quadrupole coupling on the resonance positions of the $\Delta m_I = 0$ transitions at both X-band and K-band are illustrated in Fig. 10 for both the Cu(acac)$_2$ and Na$_4$CoPTS systems. As usual, second-order hyperfine interaction shifts the resonances to lower fields and results in a monotonic decrease in the spacing of the hyperfine components with decreasing field. These second-order effects are of the order of 5–10 G at X-band, and become almost negligible at K-band. The quadrupole term, which is independent of the magnetic field, involves odd powers of m_I, and hence increases the separation between hyperfine components of the same $|m_I|$. In the Cu system the quadrupole shifts are equal in magnitude for the four hyperfine components. In the case of Na$_4$CoPTS the shifts are larger for the $m_I = \pm\frac{3}{2}$, $\pm\frac{5}{2}$ than for the $\pm\frac{1}{2}$, $\pm\frac{7}{2}$ hyperfine components. In both systems investigated the quadrupole

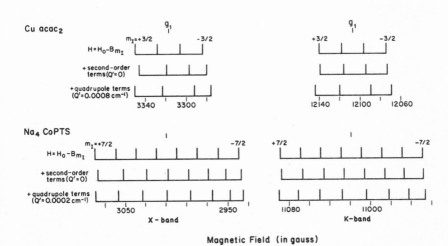

Fig. 10. The effects of second-order hyperfine interaction and quadrupole coupling on the calculated resonance positions of the hyperfine components of Cu(acac)$_2$ and Na$_4$CoPTS in the perpendicular limit at both X-band and K-band.

effects are comparable in magnitude to the effects arising from second-order hyperfine interaction at X-band. The two types of second-order correction therefore reinforce each other for the hyperfine components appearing at the low-field end of the X-band spectrum, and they tend to cancel for those hyperfine components at the high-field end. As a result, the spectrum appears considerably more asymmetrical than what one might expect on the basis of hyperfine interaction alone. At K-band the second-order corrections due to hyperfine interaction are small, and the spectrum remains fairly symmetrical even after quadrupole effects are included. However, it differs from a simple first-order hyperfine spectrum in that the hyperfine components are not equally spaced, but are more widely separated between the inner components than the outer components. In the experimentally observed spectrum these general features are, of course, masked by "extra" absorptions, rhombic components, and to some extent by $\Delta m_I > 0$ transitions. At K-band "extra" absorptions occur less frequently and the effects due to a small rhombic asymmetry become more pronounced. Both of these considerations emphasize the advantages of observing EPR spectra of powder samples at K-band.

The relative roles of hyperfine interaction, quadrupole coupling, and Zeeman interaction at both X-band and K-band frequencies are clearly illustrated in the spectra observed for the $Cu(acac)_2$ system. In addition to quadrupole interaction, a small rhombic asymmetry was readily ascertained by a comparison of the X-band and K-band spectra in the perpendicular region. In the case of Na_4CoPTS the spectral region near $H_\perp^0$ was not well-resolved at either X-band or K-band. However, the presence of a small quadrupole interaction was demonstrated, and a comparison of the results obtained at the two frequencies excluded the possibility of a significant rhombic distortion ($g_1 - g_2 < 0.005$). The lack of resolution in this spectral region is presumably due to the large number of overlapping $\Delta m_I = 0$ and $\Delta m_I > 0$ transitions and to the large intrinsic widths of the individual transitions.

The large absorption intensities observed at $\theta = 90°$ are the result of the $(1/\cos \theta)$ factor in dN/dH in Eq. (8). In addition, Eq. (5) predicts a sharp absorption wherever $d\theta/dH$ goes to infinity. The strong "extra" absorptions observed in the spectrum of $Cu(acac)_2$ have their origin in these angular anomalies. These angular anomalies result from the large anisotropy in the nuclear hyperfine interaction, and their occurrence becomes less frequent at K-band because of the increased importance of the electronic Zeeman energy relative to the hyperfine interaction at the higher magnetic field. The intensities of the $\Delta m_I > 0$ transitions are found to be almost entirely contributed by molecules with symmetry axes oriented between $\theta = 70°$ and $\theta = 90°$ relative to the applied field. Our calculations indicate that the intensities of these transitions are quite significant (see Figs. 2, 3, and 8), and that the appreciable intensity factors are largely responsible for their intensities. In the case of $Cu(acac)_2$ $(Q'/B) = 0.4$, and the relatively large Q' contributes to the intensity factor. The quadrupole interaction is relatively smaller in Na_4CoPTS. However, because of the large nuclear spin of ^{59}Co, the I^2 dependence can outweigh the small $(Q'/B)^2$ ratio in the intensity factor for many of the $\Delta m_I > 0$ transitions. From Fig. 6 we note that the angular dependences of the resonance fields for these transitions are generally quite comparable to those for the $\Delta m_I = 0$ transitions.

The analysis of the parallel spectrum is straightforward in both cases. As mentioned earlier, quadrupole interaction does not contribute to field positions of the resonances in this limit. Since B is not large, the second-order corrections due to hyperfine interaction are small, and amount to no more than 2 G at X-band. The use of Bleaney's "average energy" approximation in the perturbation energy denominators should therefore introduce little error in the analysis, even though the resonance fields of the hyperfine components are found as much as 300–400 G from $H_{\parallel}^0$ because of the large hyperfine interaction A in the parallel orientation. Despite the large second-order corrections in the perpendicular limit, which amount to as much as 10 G at X-band, the "average energy" approximation should also introduce errors of less than 1 G in the calculations of the perpendicular spectrum, since the resonance fields of the transitions are all confined within 100 G of $H_{\perp}^0$ because of the small hyperfine interaction B in this direction.

The equations employed in this work for the calculation of the resonance field positions and their angular dependences are based upon an axial spin Hamiltonian. Although the $Cu(acac)_2$ complex does not possess perfect axial symmetry, the use of these equations should introduce little error, since the rhombic asymmetry is small. In particular, the angular dependences of the resonance fields should be adequately approximated, since the distortion from axial symmetry is primarily manifested in the Zeeman interaction, which results in displacement of the resonance fields for different azimuthal angles ϕ without significantly affecting their dependences on θ. The deviation of the hyperfine interaction from axial symmetry cited in Table I is well within experimental error, and of little consequence. It should be noted that the spin Hamiltonian for rhombic asymmetry also contains the term $Q''(I_x^2 - I_y^2)$. However, since the rhombic asymmetry is small, this contribution can be ignored.[21] The value of Q' obtained in this work is in good agreement with that obtained from single-crystal studies.[1]

Since no data on the ^{59}Co quadrupole coupling constant in cobalt phthalocyanines are available for comparison with our measured value of Q' for Na_4CoPTS, a value was estimated from the Q' of ^{63}Cu in similar Cu complexes. From single-crystal studies of CuPc in H_2Pc a ^{63}Cu quadrupole coupling constant of 0.0006 cm^{-1} has been reported.[22] Q' is defined by[10]

$$Q' = [3eQ/4I(2I - 1)](\partial^2 V/\partial z^2) \qquad (14)$$

where e is the proton charge, Q the quadrupole moment of the nucleus, and $(\partial^2 V/\partial z^2)$ the electric field gradient at the nucleus along the axis of symmetry. Substituting the values of Q and I for ^{63}Cu and ^{59}Co,[23] a relationship may be derived for the ratio of the values of Q' for Cu and Co:

$$Q'(^{63}Cu)/Q'(^{59}Co) = -2.8\frac{(\partial^2 V/\partial z^2)_{Cu}}{(\partial^2 V/\partial z^2)_{Co}} \tag{15}$$

The ratio of the field gradients may be estimated by evaluating the integrals

$$\langle \Psi|(3z^2 - r^2)/r^5|\Psi \rangle$$

The field gradients were assumed to be dominated by the d electrons of the transition-metal ion, and the effects of covalency and of distortion of the $3d$-electron distribution by the ligands were ignored. The unpaired electron in CuPc is most probably in a $d_{x^2-y^2}$-orbital directed toward the coordinating nitrogens. In the case of CoPc the electronic structure assumed was $(d_{xy})^2(d_{xz}, d_{yz})^4(d_{z^2})^1$. Slater functions for the atomic d orbitals were used, and values for $\langle r^{-3} \rangle_{3d}$ were taken from Freeman and Watson.[24] This calculation indicated a value of $+1.3$ for the ratio of the field gradients. From Eq. (15), then, we obtain an estimated value of -0.00016 cm^{-1} for $Q'(^{59}Co)$. This value is in good agreement with our results. The sign of Q' was not determined by our experiments, and the sign indicated in Table I is based on these calculations.

An examination of the ^{63}Cu hyperfine interaction constants of $Cu(acac)_2$ in various media summarized in Table I suggests a possible explanation of the temperature dependence of A noted earlier in this chapter. The most plausible interpretation of these observations is axial perturbation of the complex by the surrounding media, resulting in changes in the isotropic hyperfine interaction. In a planar complex such as $Cu(acac)_2$ such environmental effects are to be expected.

In the Na_4CoPTS spectrum the reason for the apparent linewidth variation with m_I in the parallel limit is difficult to ascertain with certainty. Equation (8) predicts a monotonic increase in the intensities with decreasing field (increasing m_I) in the ratios $1.0:1.1:1.3:1.5:1.7:2.1:2.6:3.3$. Although this trend should be reflected to some extent in the derivative spectrum, it is expected to lead only to some variations in the signal heights without affecting the linewidths appreciably. A more important factor in determining the linewidths is the different angular dependences of the resonance field of each hyperfine component near $\theta = 0°$, which results in different variations of the absorption intensities with H for each parallel component. The resonance position of the $m_I = -\frac{7}{2}$ component changes most rapidly with θ, and hence the width of this resonance should be broadest, as observed. This factor would also readily account for the narrower linewidths observed for the $m_I = +\frac{1}{2}$, $+\frac{3}{2}$ components relative to the widths of the $m_I = -\frac{1}{2}$, $-\frac{3}{2}$ components. Another important contribution to the resonance widths is thought to be inhomogeneous broadening due to local variations in A. Since A, as noted for $Cu(acac)_2$, can be very sensitive to the medium, it seems reasonable to suggest that the random axial environments in frozen solutions may lead to small static variations in A for different molecules in the sample. This inhomogeneous broadening would be magnified in the outer hyperfine components and would yield the greater apparent broadening observed for these resonances. With the unpaired electron in a d_{z^2} orbital, this effect should be particularly pronounced in Na_4CoPTS. Comparison of our spectrum in DMSO and the spectra of CoPc previously reported in H_2SO_4 [6,8] with the spectrum of CoPc in ZnPc,[25] which exhibits a similar but much less pronounced variation in the linewidths, lends support to this interpretation.

CONCLUSIONS

A second-order theory, including quadrupole effects and $\Delta m_I > 0$ transitions, has been developed for the calculation of EPR spectra of polycrystalline samples of $S = \frac{1}{2}$ transition-metal ion complexes possessing axial symmetry. This theory has been applied successfully toward the analysis of the EPR spectrum of $Cu(acac)_2$ in $Pd(acac)_2$ and that of Na_4CoPTS in DMSO. The results of this study indicate that where quadrupole interaction exists it can have important effects on the EPR spectrum, and any analysis of the data without its consideration can lead to erroneous results. The advantages of obtaining spectra at two frequencies for unambiguous determination of spin-Hamiltonian parameters cannot be overemphasized.

ACKNOWLEDGMENTS

This work was supported in part by Grant GM-13212-03 from the National Institute of General Medical Sciences, US Public Health Service, and by a Public Health Service Postdoctoral Fellowship (1-F2-GM-37) from the National Institute of General Medical Sciences to L.D.R. We are indebted to Mr. Lahmer Lynds for his assistance in instrumentation.

REFERENCES

1. A. H. Maki and B. R. McGarvey, *J. Chem. Phys.* **29**, 31 (1958).
2. J. D. Swalen and H. M. Gladney, *IBM J. Res. Dev.* **8**, 515 (1964).
3. T. S. Johnston and H. G. Hecht, *J. Mol. Spectry.* **17**, 98 (1965).
4. R. Neiman and D. Kivelson, *J. Chem. Phys.* **35**, 156 (1961).
5. H. R. Gersmann and J. D. Swalen, *J. Chem. Phys.* **36**, 3221 (1962).
6. T. Vänngård and R. Aasa, in: *Paramagnetic Resonance*, Vol. II, W. Low, ed. (Academic Press, New York, 1963), p. 509.
7. L. D. Rollmann and R. T. Iwamoto, *J. Am. Chem. Soc.* **90**, 1455 (1968).
8. J. M. Assour, *J. Am. Chem. Soc.* **87**, 4701 (1965).
9. A. Abragam and M. H. L. Pryce, *Proc. Roy. Soc. (London)* **A205**, 135 (1951).
10. B. Bleaney, *Phil. Mag.* **42**, 441 (1951).
11. B. Bleaney, *Proc. Phys. Soc. (London)* **A63**, 407 (1950).
12. L. S. Singer, *J. Chem. Phys.* **23**, 379 (1955).
13. R. H. Sands, *Phys. Rev.* **99**, 1222 (1955).
14. B. Bleaney, *Proc. Phys. Soc. (London)* **A75**, 621 (1960).
15. J. W. Searl, R. C. Smith, and S. J. Wyard, *Proc. Phys. Soc. (London)* **A78**, 1174 (1961).

16. F. K. Kneubühl, *J. Chem. Phys.* **33**, 1074 (1960).
17. W. C. Fernelius and B. E. Bryant, in: *Inorganic Syntheses*, Vol. V, T. Moeller, ed. (McGraw-Hill Book Co., New York, 1957), p. 105.
18. B. Bleaney, K. D. Bowers, and D. J. E. Ingram, *Proc. Phys. Soc. (London)* **A64**, 758 (1951).
19. B. R. McGarvey, *J. Phys. Chem.* **71**, 51 (1967).
20. N. Kataoka and H. Kon, *J. Am. Chem. Soc.* **90**, 2978 (1968).
21. T. S. Piper and R. L. Belford, *Mol. Phys.* **5**, 169 (1962).
22. S. E. Harrison and J. M. Assour, *J. Chem. Phys.* **40**, 365 (1964).
23. K. Lee and W. A. Anderson, *A Table of Nuclear Spins, Moments, and Magnetic Resonance Frequencies* (Varian Associates, Palo Alto, California, 1967).
24. A. J. Freeman and R. E. Watson, in: *Magnetism*, Vol. IIA, G. T. Rado and H. Suhl, eds. (Academic Press, New York, 1965), p. 167.
25. J. M. Assour and W. K. Kahn, *J. Am. Chem. Soc.* **87**, 207 (1965).

Subject Index